Improving the Sustainable Development Goals

Improving the Sustainable Development Goals evaluates the Global Goals (Agenda 2030) by looking at their design and how they relate to theories of economic development. Adopted unanimously by the member states of the United Nations (UN) in 2015, the goals are remarkable for the global commitment on a set of targets to reach by 2030, but also for the lack of a strategy of implementation. The choice of appropriate action is handed over to individual governments, some of which are limited by their lack of resources.

This book explores how implementation of the sustainable development goals (SDGs) can be improved, especially in developing countries. The content, strengths and weaknesses of the SDGs are critically examined, alongside their relationship to ongoing academic research. The authors also investigate the actions of governments over the past three years by looking at the national strategies they have presented at annual meetings of the UN High-Level Political Forum.

Improving the Sustainable Development Goals takes a critical but constructive approach, pointing out risks as well as possible remedies. The SDGs are seen as an opportunity for a global conversation on what works in solving some fundamental problems relating to poverty and environmental degradation. With the inclusion of a chapter by Tobias Ogweno, former member of Kenya's UN mission, this book will appeal to all those who are interested in policy analysis with a focus on development issues.

Lars Niklasson is a researcher at the Swedish Institute for European Policy Studies (SIEPS), and a professor of political science at Linköping University, Sweden. His research interest is European Union (EU)–Africa relations and the global leadership role of the EU. He has taught international political economy and comparative politics, with a focus on global challenges and governance. Niklasson has been a visiting teacher at the University of Nairobi, Kenya.

Routledge Focus on Environment and Sustainability

The Greening of US Free Trade Agreements
From NAFTA to the Present Day
Linda J. Allen

Indigenous Sacred Natural Sites and Spiritual Governance
The Legal Case for Juristic Personhood
John Studley

Environmental Communication Among Minority Populations
Edited by Bruno Takahashi and Sonny Rosenthal

Solar Energy, Mini-grids and Sustainable Electricity Access
Practical Experiences, Lessons and Solutions from Senegal
Kirsten Ulsrud, Charles Muchunku, Debajit Palit and Gathu Kirubi

Climate Change, Politics and the Press in Ireland
David Robbins

Productivity and Innovation in SMEs
Creating Competitive Advantage in Singapore and Southeast Asia
Azad Bali, Peter McKiernan, Christopher Vas and Peter Waring

Climate Adaptation Finance and Investment in California
Jesse M. Keenan

Negotiating the Environment: Civil Society, Globalisation and the UN
Lauren E. Eastwood

For more information about this series, please visit: https://www.routledge.com/Routledge-Focus-on-Environment-and-Sustainability/book-series/RFES

Improving the Sustainable Development Goals

Strategies and the Governance Challenge

Lars Niklasson

Routledge
Taylor & Francis Group

LONDON AND NEW YORK

from Routledge

First published 2019
by Routledge
2 Park Square, Milton Park, Abingdon, Oxon OX14 4RN

and by Routledge
52 Vanderbilt Avenue, New York, NY 10017

Routledge is an imprint of the Taylor & Francis Group, an informa business

British Library Cataloguing-in-Publication Data
A catalogue record for this book is available from the British Library

Library of Congress Cataloging-in-Publication Data
A catalog record for this book has been requested

ISBN: 978-0-367-14210-0 (hbk)
ISBN: 978-0-429-03170-0 (ebk)

Typeset in Times New Roman
by Apex CoVantage, LLC

Contents

Figures

Preface

It is with great pleasure we have completed this joint project. The sustainable development goals (SDGs) are a very interesting global policy to deal with some of the most important challenges for humanity and the planet, with potential implications for the future. They bring a promise and aspiration of a responsible and equitable prosperity for all mankind, a peaceful co-existence of nations and an environment that supports all forms of life. The goals also demand hard work and forward-looking, yet practical strategies from governments, organizations and individuals. It is our pleasure to participate in the global conversation on how to strengthen this important and complex agenda.

Our conversation on the SDGs began in Kenya in 2017. Lars Niklasson was a visiting teacher at the University of Nairobi, while Tobias Ogweno was a diplomat at the Kenya Mission to the UN Offices at Nairobi. Tobias was the insider officer who worked under Ambassador Macharia Kamau, co-chair of the Open Working Group (OWG), which formulated the SDGs. Lars was the interested outsider, trying to understand the SDGs. Both of us are curious to see the bigger picture and contribute to the further development of national strategies.

The thinking on these matters have joined together in this book. Lars has written the theoretical parts, while Tobias has contributed a chapter on what goes on nationally. We think this mix will help the readers understand some of the challenges ahead and to think outside of the box. We think the SDGs should be taken seriously, including a serious debate on how they can be implemented nationally.

Lars Niklasson and Tobias Ogweno
Uppsala and Nairobi, January 2019

Introduction

This book suggests ways to understand and improve the implementation of the sustainable development goals (SDGs), also known as Global Goals or Agenda 2030, which were adopted unanimously by the member states of the UN in 2015. The book will do this by looking at the design of the goals, how they relate to theories of economic and political development more generally and how the goals are implemented nationally in 2016–18.

The goals are remarkable for the global commitment on a set of ambitions to reach by 2030, but also for the lack of a strategy of implementation. The choice of appropriate action is handed over to individual governments, some of which are limited by a lack of resources and other shortcomings.

The ambition of the book is to discuss how strategies for the implementation of the SDGs – especially in developing countries – can be designed. It will start from a discussion about what the SDGs are, how they were chosen and what some of the strengths and weaknesses of the SDGs are. It will relate the goals to ongoing research on relevant topics to see how the social sciences can contribute to a fuller understanding of what lies ahead.

The SDGs are treated as a case of a more general discussion about how policies can be implemented under difficult circumstances. The book advances the idea that governments can improve their policies over time, especially if they see their plans as hypotheses about what works in promoting sustainable development. To do so, they need to identify drivers and barriers to sustainable development.

The 17 goals and their specific targets are analyzed to find an inherent strategy, a statement on causal relations with drivers and barriers to achieve the goals. This strategy is subsequently related to a variety of perspectives in the research on development and sustainability challenges. The overview of research will provide a context for the discussion about the implementation of the SDGs. It will highlight the importance of thinking about ends and means, in particular, the role of states and markets in pursuing poverty reduction and other goals.

A key challenge in many poor countries is to achieve what is called good governance, i.e. an efficient administration (a state) to make decisions and implement them. The lack of a well-functioning organization may be the most important of all challenges since it makes other implementation more difficult. The book will analyze the ongoing debate on whether good governance must come first or if it is an effect of other kinds of development.

Lastly, the book investigates what governments have done over three years (2016–18) by looking at the national strategies they have presented at the yearly meetings of the UN High-Level Political Forum (HLPF). This highlights some major concerns and some lessons from the process so far.

The chapters explore the following:

- The design of the goals
- The problem with goals as policy instruments
- Hidden elements of a strategy
- How the goals relate to theories of development
- The role of good governance
- The steps governments have taken 2016–18
- Conclusions on risks and opportunities

The book takes a critical and constructive approach, pointing out risks as well as possible remedies. The SDGs are seen as an opportunity for a global conversation on what works in solving some fundamental problems relating to poverty and environmental degradation. This conversation can be improved by linking it to ongoing research on issues such as state-building and developmental states.

The book is unique in its focus on implementation and its effort to build on ongoing research in comparative politics and international political economy (IPE). There is an ambition to connect the existing research on global goals with more general research on strategies for development. The book contributes to the discussion about the SDGs by showing some problems with the design and implementation of the goals. It points out tensions which need to be discussed and handled by governments and citizens.

The book will help the reader see the bigger picture of what it takes to implement sustainable development, especially in developing countries. The book is accessible, trying to explain theories and concepts in plain language. The intended audience is professionals in the development field, involved in the implementation of the SDGs, as well as students who take courses in development policy and other areas where the implementation of the SDGs is discussed. It will also be of interest to scholars (politics, economics, sociology, global studies, etc.) who are interested in policy analysis with a focus on development issues.

The book makes five contributions to ongoing research. One is the interpretation of the SDGs as a learning policy rather than merely a set of goals. A second is the identification of an inherent strategy in the SDGs. A third is the discussion about alternative strategies with the help of a framework of perspectives from IPE. A fourth is a discussion about the importance of good governance as a foundation for sustainable development, starting from recent research which argues to the contrary, that good governance isn't important. A fifth is a discussion about experiences from the national implementation of the SDGs during the three first years (2016–18).

Overview

The first chapter will introduce the 17 SDGs by describing them and pointing out some of their key characteristics, like their lack of explicit strategies and their unclear relation to other global policies. It will discuss the concept of sustainable development as well as what is unique about the SDGs as a form of global governance.

Chapter 2 will elaborate on the ambition of the book. It will also give an overview of existing literature to point out the research gap this book aims to fill. It places the book within the boundaries of comparative politics, IPE, policy analysis and economics.

The theoretical perspective is presented in Chapter 3, where the concept of implementation is discussed. The academic debate is focussed on the concept of goals, while this book suggests that the concept of a learning policy provides a practical way forward. A learning policy is a policy by trial and error, starting from hypotheses and revising them when evidence suggests so. This makes it necessary to think about causality, i.e. in terms of drivers and barriers for sustainable development.

Chapter 4 will provide a more detailed overview of the 17 goals, including the 169 targets and some of their indicators (measurements). The chapter will discuss how the SDGs can be interpreted as a strategy, i.e. a hypothesis on what to do to bring about the desired outcomes.

Chapter 5 will elaborate on the hypothesis implied in the SDGs by giving a general overview of drivers and barriers, where alternative perspectives are highlighted. The chapter looks at theories about economic development and how economic development relates to social and environmental development. It also looks at theories about political development (good governance).

Chapter 6 digs deeper into the role of governance (the state) by testing recent research which argues, contrary to the dominant view, that economic development is more important than establishing good governance. This analysis brings out some key causal relations as well as some key values in the debate on sustainable development.

Chapter 7, written by Tobias Ogweno, discusses the experiences of the national implementation of the SDGs during the first three years – i.e. 2016–18. It focuses on common experiences and a few examples across the great variety of national activities. The analysis begins by setting out what is different with the SDGs from the previous MDGs. It ends by discussing the popular support for the SDGs.

The conclusions in Chapter 8 will discuss what can be learnt from the theoretical chapters and offer a discussion about good governance and national examples.

1 Global goals in search of strategies

We live in a time of increasing global commitments to deal with common challenges. In 2015, the nations of the world made two major agreements. In September 2015, they agreed to a set of 17 sustainable development goals (SDGs), also referred to as Global Goals or Agenda 2030 (UN 2015a). In December of the same year, they signed the Paris Agreement for action against climate change (UN 2015b). The two agreements are linked to each other, and they were both made within frameworks provided by the United Nations (UN). The SDGs came about as a replacement of the previous Millennium Development Goals (MDGs), while the Paris Agreement was made within the UN Framework Convention on Climate Change (UNFCCC). Both processes trace their history to the global meetings in Stockholm 1972, Rio 1992 and other global meetings organized by the UN.

It is remarkable that the agreements were made but, at the same time, they are surprisingly weak. Agreement came at the price of very unspecific commitments for the individual signatories. There are no ways to sanction individual governments, except for "naming and shaming". How strong this will turn out to be depends on the level of popular support for the agreements and the demands on national governments by their citizens, in democracies as well as under authoritarian regimes.

The agreements are examples of a more general situation in international politics. One aspect is the question of finding out what to do: How can broad agreements be implemented by national governments? Another aspect is the problem of compliance: Will national governments comply with the agreements, or will they look for ways to avoid doing what they have committed to? This book will treat the SDGs as a case of policy implementation, where the basic question is what governments can and should do to implement such a grand but vague concept as sustainable development.

A global agreement

On September 25, 2015, 193 heads of state and government agreed to a set of 17 SDGs as part of a UN declaration with the impressive title "Transforming Our World: The 2030 Agenda for Sustainable Development" (UN 2015a). The declaration is a global commitment to deal with very fundamental problems related to poverty, health, education and the environment. It is a commitment which almost everyone around the world can agree with, even though it may require changes in our lifestyles.

On the one hand, the global goals are a remarkable success, in the sense that there is actually a global agreement on a challenging ambition. On the other hand, the goals don't say very much about what must be done, by whom and when. Some see this lack of an explicit strategy as a strength, leaving room for national adaptation, but there is also a risk that compliance will be weak and that difficult issues will be avoided. There is now a need to produce national strategies, which means there is also a need to elaborate on the barriers and drivers, which the national strategies need to take into account. The purpose of this book is to contribute to this elaboration.

The complexity of the goals makes them interesting to study. There are big challenges in the design of strategies as well as in their implementation. Some of the challenges have to do with tensions within and among the goals themselves, while other challenges have to do with weak instruments to bring about the desired state of affairs.

A global policy for sustainable development raises hopes about such diverse ambitions as avoiding global warming, maintaining biodiversity and ensuring good living conditions for all. There are tensions and risks in these goals – for example, the conflict between raising the living standard of the poor and reducing the impact of humans on the environment. It seems that economic growth is a necessity in the first case and a fundamental problem in the second case. How can they be reconciled?

A further risk is that important trade-offs are made by distant elites, which could trigger opposition and resistance among populations across the globe. The global agenda might lead to a downgrading of democracy and a loss of legitimacy. At the same time, there is some hope that the need for joint action across the globe may lead to a greater appreciation of democracy as a way to work out acceptable compromises, taking into account the interests of all nations and their citizens. The SDGs touch upon the complex issue of what global governance should be, in terms of content as well as form of making collective decisions (Harman & Williams 2013).

Underneath the SDGs are important questions about the way we understand the global challenges and the way we think about their solutions. We live in times of remarkable economic development and a drastic reduction of poverty in countries like India and China. We also live with environmental degradation and the threat of escalating climate change. The contradictory images can be scrutinized with the help of theoretical perspectives from the social sciences to show what we have to choose between.

This chapter will give an introduction to what the goals are and what some of their problems are. The following chapters will elaborate on the need for research and what the research contribution is.

The global agreement at the UN was the latest in a long series of global summits and policy statements to deal with two of the largest challenges facing humanity, poverty and environmental degradation. The first high-level conference was held in Stockholm in 1972, followed by conferences held in Rio (1992), Johannesburg (2002) and Rio again (2012).

The SDGs are to a large extent an elaboration of the previous MDGs, which were in place for the years 2000–15. Some of the differences are that there are now more goals, that they apply to more countries and that the national governments have been more involved in the design of the new goals. I will come back to the differences between the SDGs and the MDGs, and why the goals were adopted in the first case.

Eradication of poverty in reach

Two circumstances contributed to the agreement on the SDGs in 2015. One was that the eradication of poverty seems to be within reach. This is based on the remarkable reduction of poverty over the past 25 years. Since 1990, the number of people in extreme poverty has been halved, from around two billion to less than one billion, counting poverty as living below $1.90 (originally $1.00) a day (UN 2015c). There is hope that this trend can continue and that all people will have better personal living conditions if good policies are pursued.

The reduction of extreme poverty was achieved while the global population grew, especially in poor countries, which makes it even more remarkable. However, it is mainly due to the development of China and India, not the result of a general improvement in all countries. This shows, on the one hand, that it is possible to transform very poor countries and bring better living conditions to a large part of the population. On the other hand, it implies questions about what other countries could do to deal with poverty and other social issues, as well as what China and India could do to reach the remaining poor and – more fundamentally – what the global community can and should do to assist.

Eliminating poverty would, of course, entail a major transformation of the world. Poverty is shorthand for a number of serious problems. It is closely related to hunger and disease, as well as to a lack of water, sanitation and basic services, such as education and health care (UN 2015c). Poverty is related to bad living conditions in the countryside as well as in city slums and refugee camps. Poor areas often lack the means for economic upgrading by farmers, such as electricity and organized transport (UN Millennium Project 2005). In the worst cases, there are also elements of discrimination on religious and other grounds, making it even more difficult for the poorest people to break out of their situation.

Environmental urgency

The other circumstance which contributed to the agreement is a sense of urgency in the environmental area, especially to deal with climate change and with environmental degradation in general. Rapidly growing countries like China and India are now being hurt by pollution, which calls for national as well as global action. Some of these issues are truly global in the sense that every polluter must take action if the goals are to be reached – for example, to avoid global warming.

The environmental area covers several specific problems (Speth & Haas 2006). One has to do with energy consumption, where fossil fuels are the major cause of climate change. There is a need to find alternatives to oil and coal before developing countries become dependent on these resources. Another major concern is biological diversity, which is a problem on land as well as in the oceans, where certain species are reduced as an effect of pollution. Ecosystems are changing in ways that create further problems.

We are reminded of these problems, for example, through so-called eco-labels, which have become commonplace as a way to inform consumers that products like clothing, flowers, fish and paper have been produced in acceptable ways, not contributing to the environmental problems. Environmentalism is becoming a major lifestyle, especially in developed countries.

Ambitious and flexible goals

Unlike previous policies, the SDGs apply to all countries equally, developed countries as well as developing countries. This underlines the message that we are all affected by the problems, while most of the specific issues vary across countries. More importantly, the resources to deal with the problems vary across countries, as well as the experiences from past policies. National governments have pursued a multitude of policies, which provide a large set of experiences to learn from.

The global agreement means that differences of opinion have been put to the side. Some of these differences will surface in the years to follow when governments design policies and prioritize action. Creativity will be needed to renew national policies and to make important trade-offs.

Important for the agreement was also the limited content of the global policy, i.e. that it is an agreement about goals and measurements but not about plans or strategies. The SDGs are on the one hand very ambitious, but on the other hand, they are leaving major issues open. They make up a flexible policy, which means that there is also a risk that the goals won't be reached. The flexibility provides an opportunity for a national variety of policies as well as for global learning about ways to approach the issues. More emphasis needs to be put on learning from experience to find out what works under various circumstances.

To achieve sustainable development

The two issues of poverty reduction and limiting environmental degradation are core elements of the concept of sustainable development, which has given a name to the SDGs (Sachs 2015). Sustainable development is one of the most commonly used terms in our times and has become a darling of the political debate, especially in developed countries, where it is often used as a synonym for environmentalism. The complexities of the concept are not always taken into account.

A political compromise

There is a long academic discussion about the problems with the concept of sustainable development, which is basically that it mixes several ambitions and even hides the need for a trade-off. One of the most critical versions of this critique describes the concept as a way to handle conflicts within the UN. Steven Bernstein (2001) argues that it was adopted to deal with the conflict between the north and the south after the Stockholm conference in 1972. To put it briefly, the north wanted to focus on environmental problems while the south wanted to focus on the problem of poverty. It was desirable for both to reach a consensus on a common goal, rather than clashing over contradictory ambitions.

The concept of sustainable development was developed in the 1980s – for example, by the World Commission on Environment and Development, the so-called Brundtland Commission (WCED 1987), named after its chair, the Norwegian prime minister Gro Harlem Brundtland. The concept was developed to provide the two sides with a shared goal (Bernstein 2001). This common goal was adopted at the UN conference in Rio 1992 and subsequently elaborated at later conferences.

Critics like Bernstein argue that this is a compromise which hides the need for debate on the fundamental issues. The key terms indicate that sustainable development has to do with the development of something in a sustainable way, but it doesn't say what should be developed or how it can be sustainable (Kates, Parris & Leiserowitz 2005). It is reasonable to expect the conflict to surface in the implementation of the SDGs.

"Sustainability", i.e. that something can be sustained, implies a long-term view, where such things as side effects are taken into consideration. "Development" is more difficult. It is easy to think of economic and social development, while environmental concerns are often seen more as restrictions on development, rather than a separate kind of development.

An early interpretation was that sustainability has to do with "intergenerational justice", preserving something for future generations (WCED 1987:8). This can apply to natural resources (the environment), but not to poverty, which is rather a lack of something. The concept of intergenerational justice also opens up a complicated discussion about what resources may be used, transformed, wasted, created or improved over time.

A later interpretation, adopted by the Johannesburg conference in 2002, is that sustainable development is shorthand for a mix of economic, social and environmental concerns (Sachs 2015:5). In this interpretation, sustainable development is to take all three aspects into account in a balanced way. However, the three concerns are different. The three dimensions can be interpreted as implying contradictory as well as complementary action. The relations between the three are important to investigate – for example, to decide if they are ends or means towards other ends. They can also be interpreted as restrictions on each other, a set of simultaneous ends. At this point, it is sufficient to note that there are tensions in the concept of sustainable development, which can also be found in the set of 17 SDGs.

The vagueness of the concept of sustainable development is key to its success as a political compromise. Political agreements are often based on compromises which leave room for interpretation. In modern terminology, the SDGs are a discourse, a socially acceptable ("constructed") set of problems and solutions. More controversial issues are avoided.

The vagueness is not only about the relation between the three main components but also about ways to deal with them. Many activists take the issues personally, especially the environmental issues. Some fear, though, that the focus on the environment will lead to a decreasing interest in reducing poverty. Others see a need for redistribution of resources, especially relating to the social issues. Still, others see an important role for technological development and commercial enterprise to create new solutions and new resources. Some of these issues are coming back in the debate on the SDGs.

There is something strongly emotional about sustainable development. Many advocates treat it as a personal commitment, like belonging to a political party or a social movement. Others find this disturbing and want to discuss the rationality of proposed actions: Will the national plans bring about the desired goals? What other values are at stake if the SDGs are pursued?

Contested issues

Ideas about synergies contributed to the agreement at the UN, i.e. that poverty and environmental degradation are related. However, the two issues of saving the environment and eliminating poverty are partly contradictory, which makes the global agreement even more remarkable than it first sounds. It is not self-evident that global leaders would commit to policies which place a burden on national governments to take costly action – for example, to protect the environment – when there is a great need to deal with poverty in many countries around the globe.

On the one hand, the two kinds of ambition (partly) clash with each other. They raise difficult questions about priorities in the use of resources, as well as about lifestyles, especially among the rich populations who consume most resources. However, rich countries also produce wealth which can be used to deal with the challenges. There is, for example, a great need for technological development to find alternative energy and new solutions for infrastructure and transport. New technologies can provide shortcuts of modernization in environmental protection, as well as more generally – for example, where money can be transferred through cell phones rather than banks, as in the Kenyan system of payments called Mpesa.

On the other hand, the message of the SDGs is that the problems are integrated – for example, where environmental problems contribute to poverty or when poverty leads to actions which degrade the environment (Sachs 2015). Furthermore, economic development and industrialization of larger parts of the world will most likely increase the burden on the environment, while hopefully also producing solutions. Hence, there is a need for integrated approaches in the implementation of the SDGs. This complexity will be investigated further in the following section.

Through 17 goals

The 17 SDGs cover a wide range of specific issues related to the broad concept of sustainable development. For simplicity, they can be organized into four categories, although they are meant to be integrated and implemented together. However, it is important to discuss each goal separately to open up for a better understanding of what the goals mean when they are taken together, even if there is a risk that the goals are misunderstood when looked at one by one. It is the combination of goals, with their potential synergies and conflicts, which makes up the SDGs.

Four goals deal primarily with narrowly economic issues, while seven goals deal with wider aspects of poverty, which can be labelled social issues. Another four goals are primarily related to the environment. The last two goals are about cross-cutting issues and the means for implementation. It should also be noted that goals 1–6 are based on the MDGs, while the rest (7–17) represent issues which were added through the SDGs.

Economic issues

The four goals that primarily focussed on the economy are shown in Figure 1.1.

The first of these goals (goal 1) is clearly a major goal, stating one of the most important ambitions of the SDGs, to end poverty. The ambition is very high to include everything and everyone. It opens up for a discussion about various forms of poverty as well as what the barriers are which make it difficult for poor people to break out of their conditions. Some barriers and drivers for change are implied in the following goals.

The other three goals primarily related to the economy (goals 8–10) indicate means to reduce poverty by increasing the wealth in the world. They use concepts from economic theory such as economic growth, innovation, industrialization, work, employment and infrastructure. The basic message is that the traditional focus on economic growth is still important, though within the limitation of sustainability stated in the same goals.

Goals 8 and 9 mention concepts which specify and put limits on the traditional economic concepts, such as "sustained, inclusive and sustainable" for economic growth and for industrialization. The concepts are central to the discussion about sustainable development and are open for diverse interpretations. A straightforward interpretation is that they refer to the ambitions

---

Number 1: End poverty in all its forms everywhere

Number 8: Promote sustained, inclusive, and sustainable economic growth; full and productive employment; and decent work for all

Number 9: Build resilient infrastructure, promote inclusive and sustainable industrialization, and foster innovation

Number 10: Reduce inequality within and among countries

Figure 1.1 SDGs primarily related to the economy

Source: UN 2015a

which are mentioned in other SDGs. In other words, the 17 SDGs as a package indicate a working definition of what sustainable development is. They are close to the concept of a "green economy", though it's not mentioned in the text (Kamau, Chasek & O'Connor 2018:181).

There is also goal 10, which addresses inequality. This is another concept open to many interpretations, which I will come back to. The problem of absolute poverty – living on under $1.90 a day – is dealt with in goal 1, while this goal on inequality is about relative poverty (and relative wealth), i.e. the difference between the rich and the poor. The goal takes a stand on the controversial issue of whether relative poverty is a problem like absolute poverty if differences in living standards are themselves a problem. The goal argues that the answer is "yes".

To summarize, the four goals emphasize economic growth and add some qualifications, such as a discussion about inequality. They also add some specifications of the means to achieve growth, such as industrialization and employment. I will discuss the contents further in Chapter 4.

Social issues

There are seven goals which address social issues (Figure 1.2). They can be seen as further specifications of important dimensions of poverty reduction.

A common theme is that the goals elaborate on dimensions of poverty which have been discussed for a long time, in terms of human development, etc. Food, water, sanitation, energy and health are keys to daily survival. Poor people spend much time on finding food, water and energy. Water and

---

Number 2: End hunger, achieve food security and improved nutrition and promote sustainable agriculture

Number 3: Ensure healthy lives and promote well-being for all at all ages

Number 4: Ensure inclusive and equitable quality education and promote lifelong learning opportunities for all

Number 5: Achieve gender equality and empower all women and girls

Number 6: Ensure availability and sustainable management of water and sanitation for all

Number 7: Ensure access to affordable, reliable, sustainable and modern energy for all

Number 11: Make cities and human settlements inclusive, safe, resilient and sustainable

---

Figure 1.2 SDGs primarily related to social issues

Source: UN 2015a

sanitation are bad for more than two billion people each (UN 2015c). Bad health is partly a consequence of these shortages. Education is important for the development of human potential as well as for developing national economies. Gender equality highlights the greater burden placed on women in many societies. The specific goal concerning cities and human settlements add another dimension to the list of problems.

The goals state issues which almost anyone can agree with, but they are silent on what it takes to reach them. One concern is how systems of health care and education can be organized. There is a variety of options open to developing countries (Rudra 2008). Another topic is how these systems can be funded. The goals presuppose taxation on a larger scale than is possible today, which in turn presupposes economic development to create surplus wealth (below). This is more controversial than the social issues themselves.

Environmental issues

There are four goals which relate to traditional environmental concerns (Figure 1.3).

Goal 12 is a bridge between economic and environmental issues. It formulates key aspects of sustainable development by placing limitations on consumption and production, which are the two fundamental dimensions of the economy. It is even more demanding in light of the expected growth of the population to 9.6 billion by 2050 (Kamau, Chasek & O'Connor 2018:189f).

The following three goals cover three main aspects of the environment: land, water and climate. Goal 13 makes a reference to the work on climate change within the UNFCCC – for example, the so-called Paris Agreement 2015 (UN 2015b). It was argued in the negotiation phase that the SDGs had to have a goal on climate change, even if there was a risk that this would limit the subsequent negotiations in Paris (Kamau, Chasek & O'Connor 2018:192ff).

Number 12: Ensure sustainable consumption and production patterns

Number 13: Take urgent action to combat climate change and its impacts

Number 14: Conserve and sustainably use the oceans, seas and marine resources for sustainable development

Number 15: Protect, restore and promote sustainable use of terrestrial ecosystems, sustainably manage forests, combat desertification and halt and reverse land degradation and halt biodiversity loss

---

Figure 1.3 SDGs primarily related to environmental issues

Source: UN 2015a

The environmental goals are similar to the social goals in that they don't mention what it takes to reach them. Some more specific issues are mentioned in the targets, but there is very little on strategies and how these goals relate to economic development, which is an essential aspect of the economic and social goals. The economic goals (noted earlier) have explicit links to the environmental goals.

Cross-cutting issues

The final goals highlight cross-cutting issues which can be interpreted as prerequisites to reach the other goals (Figure 1.4).

Goal 16 is interesting because it widens the range of factors which are relevant to take into consideration. Economic, social and environmental issues are part of the concept of sustainable development, while this goal adds a concern for peace, justice and institutions. The latter three concepts widen the scope by bringing in questions about politics and governments. They hint at factors which are important in order to achieve the other goals, such as regulation and taxation. In other words, peace, justice and institutions can be seen as prerequisites for the other SDGs.

It was suggested, but not agreed on, in the preparations that these issues should form a fourth dimension of sustainable development (Kamau, Chasek & O'Connor 2018:201ff).

It is easy to understand that "peaceful" is added to the list of goals since conflicts are a problem in many parts of the world and make it difficult to deal with the other goals. Justice and institutions refer to the role of governments in society. They imply that problems such as corruption and dictatorships need to be dealt with. Furthermore, to talk about "effective, accountable and inclusive institutions" is almost to say that democracy is a necessary condition. It can also refer to the institutions (regulation) necessary for a well-functioning market economy. "Justice for all" highlights the impartiality which is central to not only the justice system but also to government bureaucracies, often referred to as the rule of law. The latter

---

Number 16: Promote peaceful and inclusive societies for sustainable development, provide access to justice for all and build effective, accountable and inclusive institutions at all levels

Number 17: Strengthen the means of implementation and revitalize the Global Partnership for Sustainable Development

---

Figure 1.4 SDGs primarily related to cross-cutting issues

Source: UN 2015a

term was controversial and could not be used in the goal itself (Kamau, Chasek & O'Connor 2018:202f). In summary, the goal indicates the importance of factors which are often referred to in the debate on development as "good governance", which I will discuss further in Chapter 6.

Goal 17 is different in the sense that it addresses issues which are related in general ways to the implementation of the SDGs, such as funding and technological development. I will elaborate on these further in Chapter 4.

Without an explicit strategy

A set of goals, like the SDGs, is a special kind of policy. The SDGs are a hybrid form of governance arrangement. They are neither a global plan to be implemented top-down, like the infamous Soviet Gosplan, nor are they just an encouragement of bottom-up initiatives. They are an open framework, with some predefined variables in the form of goals, targets and indicators for measurement. It is up to the national governments to apply the framework to their specific contexts, i.e. to "connect the dots".

Weak on strategies for implementation

The openness regarding the implementation is a strength as well as a weakness, especially since there is no general agreement on what specifically a national strategy for sustainable development should look like. For the national governments, the goals allow great flexibility but demand a great effort in terms of designing policies, making priorities and finding instruments to bring about the desired effects. They also need to learn from experience to improve policies over time.

The goals state a number of problems to be solved individually in each country. The strategies are up to individual governments to design within the global framework. A relevant question is, therefore, if national governments have sufficient capacity to select and implement plans and strategies. Governments differ in their capacity for political action. Countries differ in the range of solutions that are available, as well as in the challenges they face.

The goals address fundamental challenges which are contested and often without obvious or proven solutions. This makes it relevant to think about implementation issues and how the global goals impact on actors and existing policies across the globe. Some of the goals are very close in content to the Paris Agreement on climate change within the UNFCCC. Other goals relate to the operations of global organizations like the International Monetary Fund (IMF), the World Bank, the International Labour Organization (ILO) and the World Trade Organization (WTO).

The vagueness of the goals is part of the problem. The goals have to be general to apply to all countries, which means that there is a need to adjust the goals to the specific situation in each country. The goals don't specify who should do what, when or how. This adds a great amount of flexibility on the one hand, but on the other hand, it means that difficult issues were postponed for later. It becomes necessary for national governments to adapt the general framework to the local circumstances. This, in turn, adds a further element of uncertainty to the global agreement: Will all countries comply with the agreement and do what it takes to reach the goals?

To some extent, the SDGs are based on previous international agreements, which means that there is sometimes another source of binding commitments. There is a discussion among legal scholars on how binding various global declarations actually are (Kim 2016). To some, the SDGs are mainly a political statement of commitment to give more attention to these issues, a response to increasing demands from populations across the globe.

How flexible are the goals?

Most of the goals state general principles which can be interpreted in a variety of ways. The document leaves many possible interpretations open, focussing on individual goals or working out a collective meaning, taking all the 17 goals into account. A common interpretation is that they should be treated holistically, as a set, meaning that some goals limit the interpretation of other goals. In other words, the SDGs are more than just a compromise on a disparate set of priorities. There is some rationality in the choice of goals. Hence, the need to take all of them into account in the development of national strategies.

Furthermore, each goal has a set of specific targets, making a total of 169 targets to be reached. These targets give focus to the goals, indicating specific problems to be dealt with. It is the set of targets which spell out the meaning of the 17 goals. It is not enough to discuss what the 17 goals mean since that would be to open up a much wider set of interpretations. The key to the SDGs is to understand what the 169 targets demand from national governments. In other words, a key question is to find out if there is an inherent strategy in the 169 targets, a coherent message to the national governments about what they are expected to do in the implementation phase. This will be discussed in Chapter 4.

A set of quantitative indicators for each goal adds even more specificity, pointing to specific aspects to be dealt with and reported on. A common risk is that only those aspects which are measurable and selected by the UN become important in the national context. There is a difficult trade-off between flexibility and rigidity in the global policy, which is about giving room for adaptation vs. making sure that at least some issues are dealt with (i.e. avoiding shirking), which I will come back to in Chapter 5.

Weak on compliance

The goals are a collective endeavour, with mutual obligations between the national governments. The SDGs are designed as a global framework of policy to be implemented nationally. They are part of the overall Agenda 2030 and supported by global commitments on funding and technical support, etc. (UN 2015a). These arrangements are important, even a sign of a strong global commitment to solving common problems.

The two problems of implementation are very different. While the national problem is to work out a reasonable strategy, the global problem is to make sure that all governments take reasonable action. The latter problem is about compliance and loyalty to the group of countries which have committed themselves to the SDGs. It may be tempting for individual countries to cheat on the agreement while letting others take a greater burden. This is the classical problem of compliance with international agreements (Hasenclever, Mayer & Rittberger 1997).

The SDGs can be compared to the great variety of governance arrangements at the global level, ranging from declarations without legal force, to supranational organizations which act independently on the basis of powers which have been handed over to them from national governments. The goals are closer to the former than the latter. The goals are weak in the sense that they are primarily based on a moral commitment among national governments, even though there are sometimes other binding agreements behind the goals.

The main mechanism to ensure compliance (commitment) by the states which agreed to the policy is group pressure. There are Voluntary National Reviews (VNRs) on how well the countries are doing, which can be the basis for praise or criticism from other states and other interested parties. In the language of international politics, this is described as "soft law", in contrast to traditional "hard law", which creates legal obligations on the signatories (Nye 2011).

The problem of compliance depends partly on the character of the goal and whether it can be reached only by joint action if the participation of all countries is necessary. The pressure for compliance is stronger if the "bad countries" spoil the situation for all. If this is not the case, the problem becomes one of bargaining to make up for those who cheat (Barrett 2007).

This, in turn, may affect the willingness of citizens to change their behaviour. Those who are sceptical of the goals may be conditional compliers – i.e. they comply on the condition that everyone else does too. For example,

Americans might think that the goals are relevant only if Europeans and the Chinese follow the goals. This is often described as the need for policies to be legitimate in the eyes of those affected. Low legitimacy can be the result of a questionable ground for making the decision – for example, when the mandate from the citizens is weak.

A reverse situation is where some citizens or countries take on greater obligations to become role models for the rest, an altruistic strategy. The concept of a "Minervian strategy" refers to a strategy where altruistic moves are taken to make self-centred actors cooperate more than they would otherwise do (Tiberghien 2013). There are generally more altruistic motives in policy areas such as environmental protection and support for the poor than there are in domestic policy areas. This indicates that citizens and, possibly, governments are willing to make sacrifices in the implementation of the SDGs.

The character of the problem to be solved through collective action differs across the goals. Climate change is affected by the actions of all, while the same isn't true about poverty reduction or environmental degradation in general. These issues have more local effects. Such differences may have an impact on the willingness to implement the goals.

And in an unclear relation to other global policies

As noted, the SDGs are not the only international policy in relation to poverty and the environment. The environmental goals make reference to other global agreements and ongoing negotiations, where the Paris Agreement on climate change is the most visible but not the only overlapping arrangement. As mentioned, there is a debate on how much of the SDGs find support in previous binding agreements such as international conventions.

There is an ambition that other global arrangements (policies, organizations) should support the implementation of the SDGs, but the reality could be that they clash with each other (Cormier 2018). It was noted under the previous MDGs that the goals created confusion within at least one international organization, the International Monetary Fund (IMF). The problem was that the instructions for the IMF weren't entirely compatible with the MDGs, which in a sense meant that the member states had given incompatible instructions to the IMF and to the UN in the form of the MDGs (Gutner 2010). Such conflicting policies can lead to a situation where there is more room for the organization to decide independently how it prioritizes actions.

The problem of a lack of coherence opens up loopholes for national governments, or even competition between the arrangements, with a corresponding option for national governments to select where they want to pursue their interests and to set up new arrangements when they don't like the old ones (Morse & Keohane 2014).

At the same time, the SDGs are an example of a new multilateral commitment – although with few binding obligations – while other multilateral arrangements are facing trouble. Comparing the SDGs to international organizations brings out what the SDGs are not and what they might have been. The counterfactual question is why the SDGs are only goals with little binding power over the signatories. Could they have become an international organization with resources to support national governments? Could there have been more powers turned over to the UN Developmental Program (UNDP) and the UN Environmental Program (UNEP)? Are the SDGs even a way to move forward while the idea of setting up a World Environmental Organization is blocked?

An even wider question about the role of the SDGs in global politics is how they relate to what is called the Liberal World Order, the set of regulations and organizations that were formed at the end of the Second World War to promote values such as democracy and market economy (Sørensen 2011). The SDGs are not only compatible with the Liberal World Order but gives a prominent role to organizations like the UN and the World Bank. Some critics would argue that the SDGs are an example of the power of Western (northern) countries to dominate the agenda. Other critics might argue to the contrary that the SDGs are only a weak attempt to gain the support of developing countries for the Liberal World Order. In this version, the Western (northern) countries need to show with stronger emphasis that they want to support the development of the economies and polities of the global south, especially since there is a widespread view that support (aid) is given in partly contradictory ways, not always focussing on the needs of the recipient countries.

But still an example of the art of the possible

A final introductory remark is that the history of the SDGs may shed some light on what to expect from their implementation. It is remarkable that the UN resolution (UN 2015a) was accepted by consensus. This creates some loyalty across governments, which gives extra strength to the implementation processes. At the same time, conflicts at the negotiation stage may indicate later problems at the implementation stage.

The consensus was arrived at after a long period of discussions and intense work by a limited number of elected representatives of national governments around the world. Other authors have documented the various phases of this work (e.g. Kamau, Chasek & O'Connor 2018).

Discussions began by the UN General Assembly in 2010 on how to follow-up the MDGs, which were in place for 2000–15. More than 1.4 million participants contributed to the process of consultations (ibid.). The

Secretary-General established a UN System Task Team in 2011. A high-level panel was formed in 2012 to make recommendations in 2013.

This topic was the central issue at the international conference in Rio 2012, where a broad decision was made to develop a new set of goals to replace the MDGs after 2015. The UN General Assembly set up an Open Working Group (OWG) to negotiate a draft. It had a set number of mandates for different parts of the world (regions of the UN). The seats were often split between two or more interested countries. The OWG managed to come up with a draft based on consensus.

It is also remarkable that the General Assembly of the UN decided to stick to the proposal by the OWG rather than open up for a full round of negotiations. A key element is that this was a political compromise to bring everyone along. The idea of a global commitment created its own momentum and the emphasis on goals rather than means made it easier to reach an agreement.

There were several difficult issues during the negotiations. Some were hinted at earlier, such as the conflict over the term "rule of law", which is mentioned only in a target but not in the goal (goal 16). Family planning was another contested issue, mainly for religious reasons. The most interesting conflict from the perspective of internal tensions in the concept of sustainable development was over industrialization, which environmentalists rejected. It was maintained after a package deal consisting of the goals 9 (industrialization), 10 (equality), 12 (consumption and production) and 16 (peace, justice and institutions; Kamau, Chasek & O'Connor 2018:203).

Another contested issue is how different the new global policy really is. One side sees the SDGs as a break with previous global development policies, especially the focus on economic dimensions, often called "neo-liberal" policies, exemplified by the infamous "Washington Consensus" (Serra & Stiglitz 2008). Here, the focus on poverty and the environment is seen as a contrast to a focus on the economy.

Others see more of continuity, where governments still have to live up to demands for balanced budgets and economically sound policies. In the latter perspective, the advantage of the new policy is that it is more attractive to a wider audience by putting greater emphasis on the problems of society, seeing the economy more as a means than an end. Furthermore, a focus on policy goals might lead to a greater interest in the underlying drivers and barriers for development, including the economic means to reach the goals.

There are other elements which are worth noting. One is the obvious link to the MDGs, which provided a starting point for the negotiations. It is the adoption of the MDGs in 2000, which is the big turning point in global development policy. The focus on poverty was new and partly in contrast to the tougher policies under the so-called Washington Consensus, which focussed on making governments pursue liberal policies, i.e. to promote democracy, market economy and sound macroeconomics.

Setting out the policy in the form of goals was new and in line with results-oriented policies pursued by many national governments. These elements made the MDGs attractive to a large audience, to make a long story short (Fukuda-Parr & Hulme 2011). This says something about which ideas (norms) dominate in this policy area and how governments (and the development policy community) like to deal with the issues.

This chapter has given an overview of what the SDGs are and what is important to bear in mind to understand the implementation of the goals. The next chapter will elaborate on the need for more research and what the contribution of the book will be.

References

Barrett, Scott 2007: *Why Cooperate? The Incentive to Supply Global Public Goods*, Oxford: Oxford University Press.

Bernstein, Steven 2001: *The Compromise of Liberal Environmentalism*, New York: Columbia University Press.

Cormier, Ben 2018: "Analyzing If and How International Organizations Contribute to the Sustainable Development Goals: Combining Power and Behavior", *Journal of Organizational Behavior*, vol. 39, pp. 545–558.

Fukuda-Parr, Sakiko & Hulme, David 2011: "International Norm Dynamics and the 'End of Poverty': Understanding the Millennium Development Goals", *Global Governance*, vol. 17, pp. 17–36.

Gutner, Tamar 2010: "When 'Doing Good' Does Not: The IMF and the Millennium Development Goals", in D. D. Avant, M. Finnemore & S. K. Sell (eds.): *Who Governs the Globe?*, Cambridge: Cambridge University Press.

Harman, Sophie & Williams, David (eds.) 2013: *Governing the World? Cases in Global Governance*, Abingdon: Routledge.

Hasenclever, Andreas, Mayer, Peter & Rittberger, Volker 1997: *Theories of International Regimes*, Cambridge: Cambridge University Press.

Kamau, Macharia, Chasek, Pamela & O'Connor, David 2018: *Transforming Multilateral Diplomacy: The Inside Story of the Sustainable Development Goals*, London: Routledge.

Kates, Robert W., Parris, Thomas M. & Leiserowitz, Anthony A. 2005: "What Is Sustainable Development? Goals, Indicators, Values and Practice", *Environment: Science and Policy for Sustainable Development*, vol. 47, no. 3, pp. 8–21.

Kim, Rakhyun E. 2016: "The Nexus between International Law and the Sustainable Development Goals", *Review of European Community & International Environmental Law*, vol. 25, no. 1, pp. 15–26.

Morse, Julia C. & Keohane, Robert O. 2014: "Contested Multilateralism", *Review of International Organizations*, vol. 9, pp. 385–412.

Nye, Jr., Joseph S. 2011: *The Future of Power*, New York: Public Affairs.

Rudra, Nita 2008: *Globalizaton and the Race to the Bottom in Developing Countries: Who Really Gets Hurt?*, Cambridge: Cambridge University Press.

Sachs, Jeffrey D. 2015: *The Age of Sustainable Development*, New York: Columbia University Press.

Serra, Narcis & Stiglitz, Joseph E. (eds.) 2008: *The Washington Consensus Reconsidered: Towards a New Global Governance*, Oxford: Oxford University Press.

Sørensen, George 2011: *A Liberal World Order in Crisis: Choosing between Imposition and Restraint*, Ithaca: Cornell University Press.

Speth, James Gustave & Haas, Peter M. 2006: *Global Environmental Governance*, Washington, DC: Island Press.

Tiberghien, Yves (ed.) 2013: *Leadership in Global Institution Building: Minerva's Rule*, Houndmills: Palgrave Macmillan.

United Nations 2015a: *Transforming Our World: The 2030 Agenda for Sustainable Development*, Resolution adopted by the General Assembly on 25 September 2015, A/Res/70/1.

United Nations 2015b: *Adoption of the Paris Agreement: Proposal by the President*, Framework Convention on Climate Change, Conference of the Parties, Twenty-First Session, FCCC/CP/2015/L.9/Rev.1.

United Nations 2015c: *The Millennium Development Goals Report 2015*, New York.

United Nations Millennium Project 2005: *Investing in Development: A Practical Plan to Achieve the Millennium Development Goals*, New York: Earthscan.

WCED, World Commission on Environment and Development ("the Brundtland Commission") 1987: *Our Common Future*, Oxford: Oxford University Press.

2   A research gap on strategies and implementation

The SDGs raise a number of questions. In this chapter, I will explain the questions I'm asking and how they relate to previous research on the SDGs and the more general issues of poverty reduction and environmental protection.

Focus on the implementation of the SDGs

There are two types of questions academic research asks about policies. One is backwards-looking and often critical, asking why a certain content was chosen, what alternative options there were and what values were selected to be promoted through this policy. This is often tied to showing who supported the policy and which interests it gave support to. A modern version looks at underlying perspectives to bring out the assumptions politicians make, consciously or unconsciously. Political science, history, sociology and critical studies of various kinds do this type of research, emphasizing such things as power relations and the use of rhetoric and norms.

Forward-looking

The other type of question is forward-looking, asking where the policy will lead, if it is based on reasonable instruments, if it will achieve the desired ends. This research has a more instrumental view of politics, taking the stated aims as given, but can nevertheless provide inputs to future policies, on the relationship between ends and means. Economics is concerned with policies and their outcomes. Political scientists ask about the role of governments in achieving desired ends. The field of comparative politics discusses general questions about governance – for example, how democracy, the rule of law and an efficient public administration contribute to social goals and how they are established in the first place (Landman 2013).

Typical forward-looking questions are: What will happen now? National leaders have agreed to the goals, but what will they do about them back home? Is the commitment strong enough? Are there strategies and instruments to achieve the goals? How will international commitments support national policies? Will the SDGs be a continuation of the MDGs in developing countries? And how will they be dealt with in developed countries, which were not part of the MDGs?

Over time this type of research will develop some backwards-looking, by investigating the links from intentions to outcomes. Typically, a number of unexpected outcomes will be produced by any policy. The contribution of implementation research is to understand what happens and why it happens, bringing in factors which have an influence on the implementation, especially factors which were overlooked during the negotiations and decision-making (Hill & Hupe 2002). An example of such contextual factors and perspectives is how climate mitigation policies in rural Zambia become part of a struggle over authority, where national agencies aim to strengthen their position with the help of the new policies while local authorities have a similar ambition (Funder, Mweemba & Nyambe 2018).

My focus is forward-looking, investigating strategies for the implementation of the SDGs. I have looked backwards in my description of the SDGs, as a way to understand some of the specific problems that lie ahead. I have given a short overview of important characteristics of the processes leading up to the SDGs in the previous chapter. However, most of my analysis will look at fundamental drivers and barriers, which are played down in the SDGs. Such drivers and barriers are discussed in the general literature on development – for example, the importance of good governance.

The topic of governance is included in the SDGs, but only in of one of the goals (goal 16). My ambition is to deal with these issues as a background condition for the achievement of all the SDGs. I will focus on strategies to achieve the SDGs and especially relate them to research on the fundamental importance of governance (institutions, state capacity, accountability, etc.) for the achievement of goals like poverty reduction, human development (health and education) and environmental protection.

My ambition is to begin an analysis of how the SDGs can function as a national policy. It is the implementation of the goals which is my interest. Since it is extremely complex and has only just begun (even though there is some continuity from the MDGs), the ambition is to set out a framework for analysis, specifying the design and inherent strategy of the SDGs (Chapter 4). My ambition is furthermore to relate the SDGs to major perspectives in the social sciences on how to deal with these issues, especially the role of "good governance" in the implementation of the SDGs (Chapters 5 and 6). Chapter 7, by Tobias Ogweno, will look at the national implementation of the SDGs. The issues are contested in the academic community and an important aspect is therefore to point out several types of contributions from academic research.

The aim of the study is, in other words, to evaluate the SDGs beforehand ("ex ante") by looking at their design and how they relate to theories about development. It is an example of policy analysis, which aims at improving our understanding of what governments can achieve and how. The ambition is to draw conclusions which can be used in other cases. This is also to say that it deals with general issues about the design of policy, the use of tools and the capacity of governments to carry out policies and to achieve desired outcomes.

Implementation studies are often conducted after a policy has been implemented, to find out what happened and to explain this by following the chain of events. An alternative is to investigate policies at the beginning to make predictions and suggestions about revisions. In either case, a description of the intended policy is a key item.

My focus is on the inner logic – the inherent strategy – of the SDGs rather than on the concept of sustainable development. There are other books which add interpretations of what it takes to reach the goals, especially the environmental ones (e.g. Sachs 2015). They downplay the alternative interpretations, which I will give large room in Chapter 5.

Connecting policy to research

My contribution is a friendly critique of the SDGs. I will point out problems with the policy, with the intention that they can be addressed by national governments. As with other highly complex policies, it is reasonable to think of the SDGs as a learning type of policy, where governments learn from experience and from research to continuously improve their policies, as a kind of implementation by trial and error. I will elaborate on this in the following chapter. The assumption is that governments are willing to learn, to improve the outcome of their policies.

My contribution is more specifically to set the implementation of the SDGs in a context of research on development, the bigger issue which the SDGs are a particular case of. Others have interpreted the SDGs as a case of environmentalism, by putting a focus on what the environmental goals demand of us. My ambition is to be more open and look at the combined concerns within the SDGs and especially how this relates to developing countries.

I will use the wider research as a foundation for a discussion about drivers and barriers for the SDGs. Here I will especially use research from comparative politics, which is the section of political science where individual countries are studied and compared to each other. One major theme in this area is to understand how states and democratic forms of governance evolve under various circumstances, which entails comparisons of successful cases over time with cases (countries) where few steps have been taken (Landman 2013).

A fully developed system of democracy may not be necessary to achieve some of the SDGs. History shows that authoritarian regimes can be successful in driving social change, for example, to stimulate economic growth (ibid.). This is one of the issues debated in the research on drivers and barriers for political change, which I will elaborate on in Chapter 5.

The debate on political development has some overlap with the economists' focus on explaining economic development, since the two types of development are often interrelated. There is a converging interest among researchers on the role of governance ("institutions") as a foundation for economic development. On the other hand, economic development has a more difficult relationship with political development. Economic development can sometimes be a driver and sometimes act as a barrier for political development – for example, where resource-rich countries either end up in civil war or with dictatorships ("the resource curse"; ibid).

Economics and comparative politics ask questions about the drivers for economic development, which often focus on political institutions (regulations, policies, etc). This implies that drivers and barriers can be found in the way political systems are organized. Hence, serious efforts to bring about sustainable development must take these issues into account.

Economic development can have positive as well as negative effects on sustainable development. Some environmentalists see the capitalist economy as the root cause of environmental degradation through overconsumption and human greed (Newell 2012). Other environmentalists see this rather as a problem with the legal system (institutions) not functioning properly, as in the fable of "the tragedy of the commons", where collectively owned resources are exploited (Andersen & Libecap 2014). A positive view of economic development would add the importance of technological development, which is partly driven by competition among major firms. This variety of perspectives will be discussed in Chapter 5.

This is a type of policy analysis based on research in comparative politics and economics. Policy analysis is a field of research concerned with the design of policies. It is forward-looking and instrumental (Howlett & Ramesh 1995). It has been criticized for taking existing policies for given, rather than investigating why particular policies were chosen. As I indicated earlier, I find both questions valid. I find it reasonable to take the SDGs seriously by investigating where they will lead – i.e. if it is likely that the design of the policy will lead to the achievement of the goals. This type of "instrumental knowledge" can be useful in the further development of policy and will contribute to the general understanding of drivers and barriers for development.

The tradition of asking instrumental questions is evident in economics as well as in comparative politics. The two share an ambition of giving advice on what works. Where they differ is in the focus and in the methodology. Economics is a theory-driven discipline, generating hypotheses in a deductive way, while comparative politics is mainly empirical, building theories inductively from data. Economics has produced elaborate theories and has become influential among policymakers, partly by being able to derive policy proposals from their theories. Critics argue that the theories are based on controversial assumptions, which are overlooked by policymakers (Rodrik 2011).

I will also use a framework for studying policies from IPE, where alternative perspectives are compared (Miller 2018). Here, the idea is that it is useful to contrast several perspectives to bring out what conflicts are about, rather than taking a particular perspective as given. The perspectives indicate a policy space of possible positions, a kind of measuring device for understanding contrasting views on the SDGs.

Africa and developing countries

I will focus on Sub-Saharan Africa in my analysis, especially in Chapter 6. This may seem odd, when one of the novelties with the SDGs is that they apply to all countries and make a point of not focussing on developing countries. There are two reasons for this analysis to be conducted differently. When it comes to drawing conclusions, it is often better to look at countries which are similar in fundamental respects. This is why comparative politics is often referred to as area studies, limiting the investigation to specific parts of the world where many factors are the same and therefore can be ignored in an analysis of causal relationships. Comparative research looks for a limited set of differences between contexts which can explain differences in outcomes (Landman 2013).

The argument for looking specifically at Africa is that it is an area of the world where several fundamental problems have to be dealt with. Hence, the greatest contribution of this research would come from giving advice on Africa. There are, of course, many other parts of the world which face similar problems, but the logic of area studies suggests that one should pick only one area.

I will touch upon an alternative way of asking the question that follows, asking how Asia and Africa differ. Development economists especially ask why Asian countries like South Korea have managed to transform themselves from poor countries to rich countries in about 50 years, while African countries, such as Nigeria, haven't (Henley 2015). This is the researchers' way of zooming in on differences in the context which can explain differences in outcomes (economic growth). However, there are many background conditions which differ and make it difficult to come to agreement on the causal relationships. Some of the drivers for economic growth are international, like the support from dominant countries.

Questions for the investigation

This reasoning boils down to three questions I want to address in the following.

The first question is, is there a way of understanding the policy (the SDGs) which gives it a fair chance of achieving its goals? In other words, if the means of implementation is a problem, could there be other approaches – such as learning over time – which would improve the workings of the policy? I will show in the next chapter that this is a reasonable starting point for the investigation.

The second question is how are the SDGs designed as a policy? What are the instruments which will be used (briefly described earlier)? Is there an inherent strategy in the 169 targets to be achieved? This is an empirical question to be analyzed in Chapter 4.

The third question is how can the SDGs take advantage of ongoing research to spell out some more elaborate hypotheses on what needs to be done – i.e. to identify drivers and barriers for the achievement of the goals? This is a very broad question which will be divided up in two parts and then answered through an overview of research. I will look at alternative views of drivers and barriers in Chapter 5, and I will focus more specifically on the topic of governance in Chapter 6.

Finally, Tobias Ogweno addresses the question of what is done by national governments to implement the SDGs in Chapter 7. He will provide some information on the context and some on the problems faced by governments. This is the beginning of an analysis of what works in the implementation of the SDGs and why it does so under various circumstances.

Are the goals intended to be taken seriously?

One problem with interpreting the SDGs is if they are meant to be taken literally or not. On the one hand, there were intense negotiations over some of the issues. The set of goals is what there was agreement on. This suggests that the implementation should stay loyal to the exact wording of the goals and targets. On the other hand, there is a need to think more freely, to interpret and point out what is taken for granted and what further issues need to be addressed. For example, there are elements that can be criticized for being weak, such as the text on fossil fuels (below).

My position is that we should take the SDGs seriously in the sense that we should start from the exact wording, try to understand what the text means and then add possible interpretations to bring out implications of what the text means. This is a traditional way of dealing with political texts, focussing on what it means (Vedung 1982).

In this case, the ground for alternative interpretations comes from the specific debates on each topic. The SDGs are in a way taken out of context, hence an alternative is to see what is meant in the specific debates on each topic. The problem is that this will most likely go beyond the intentions of at least some of the parties of the negotiations on the SDGs.

I will do a version of this, where I will discuss alternative interpretations from a set of positions from the social sciences. As I will explain in Chapter 5, the ambition is to present the major perspectives, although in a pure and somewhat simplified form. There are a large number of specific positions in the debate on the SDGs, and one way to make sense of them is to identify opposites, which will help us understand what the fundamental debates are about.

Another alternative is to bring in the legal context. The SDGs sound as if they are a stand-alone document, but in fact, they make a statement on issues which are partly regulated through international law and dealt with by international organizations (Cormier 2018). Thinking like a lawyer, it makes sense to describe a context of international treaties on relevant topics. One could argue that this was an important background for the parties of the negotiations. The SDGs are weaker than the older and more exact legal documents, though one can argue about the strength of the legal force of the SDGs. The documents adopted by the UN General Assembly have no legal force but some moral force, especially if adopted unanimously by all member states. The other documents have more of a legal force.

It is difficult to find all relevant legal documents in relation to the SDGs, but an effort has been made through a Canadian organization known as the Centre for International Sustainable Development Law, together with the UNEP.

Problems with the existing literature

My questions have only been asked to a limited extent. Academic scrutiny has mainly dealt with the design of the SDGs and the previous MDGs, and with reporting on what has happened in various countries. Such analysis mainly stays on the surface, dealing with description and some limited criticism. At most, it treats the goals as a modern discourse on development, pointing out that there are alternative ways to think about what needs to be done. The research has only to a small extent dealt with the importance of drivers and barriers, which is discussed elsewhere – for example, in comparative politics.

To put it more forcefully, the research on the SDGs and the MDGs could put more focus on the causal assumptions of the policies. The goals can to a larger extent be related to the academic literature on the challenges faced by developing countries. My ambition is to bring in this literature to show what is either ignored or taken for granted in the design of the policies. As I said before, I do it with the intention of improving the implementation of the SDGs.

A volume which addresses some of the causal assumptions is *From Millennium Development Goals to Sustainable Development Goals. Rethinking African Development* edited by Kobena Hanson, Korbla Puplampu and Timothy Shaw (2018). It gives a nice overview of some of the relevant issues and covers several interesting topics, such as the problems in the development of agriculture and the importance of good governance for development in general. While it mentions relevant research, the number of factors and countries makes it difficult to draw conclusions about what works, where and when. It is an overview rather than an in-depth study to identify drivers and barriers under particular circumstances.

The overview of a selection of relevant factors is in contrast with comparative politics, which aims to draw conclusions about empirical (causal) relationships through in-depth descriptions of cases, often by comparing cases with interesting (minor) differences (Landman 2013). Such comparisons are carried out by means of structured analysis, which can be done by the use of statistics, if data is available (ibid.).

There is an older tradition in political science where differences of opinion are what drives the discipline. In the old tradition, one should look for opposite positions in the debate on development, extremes which can be contrasted and cover the range of hypotheses on the topic (Vedung 1982). In this tradition, the main question would be to find alternative views on what is needed to succeed with the goals. Typically, liberal and Marxist perspectives would be contrasted, sometimes with a statist (nationalist) position in between, as I will do in Chapter 5. These extremes are the far corners of a triangle of pure positions, which provides a map of possible instruments for policymakers to use or avoid (a "policy space"). They point out alternative instruments to reach the SDGs.

Sustainable development

There is a general literature on sustainable development, which is relevant but difficult to use. A problem is that the concepts of the environmentalist literature are often vague. It seems that the research in this area is still in the process of coming to agreement on what concepts such as "sustainability" and "resilience" mean and how they can be implemented. Examples of this vagueness are found in *Resilience, Development and Global Change* by Katrina Brown (2016), which is mainly a discussion on how to understand the concept of resilience. In the previous chapter, I mentioned the view that there are inherent contradictions in the concept of sustainable development (Bernstein 2001).

Another type of literature in this field is similar to my own investigation. This is the literature which defines and compares positions, indicating broad world views and how their proponents think about ends and means. One example is *Paths to a Green World. The Political Economy of the Global Environment* by Jennifer Clapp and Peter Dauvergne (2011), which identifies four basic perspectives in the debate on sustainable development: market liberalism, institutionalism, bioenvironmentalism and social greens. Three of these perspectives have great similarities with liberalism, statism and Marxism, which I will come back to.

Theoretical perspectives

The literature focussed on the SDGs are basically of two kinds: theoretical or empirical. The theoretical type focuses on the specific design of the global policy in terms of goals. The concept of goals is in focus, and the comparison is with other ways of designing policies. These books discuss some issues of implementation, but generally not the possibility of a learning policy, which is my contribution (next chapter).

Examples are *Millennium Development Goals. Ideas, Interests and Influence* by Sakiko Fukuda-Parr (2017) and the volume *Governing through Goals. Sustainable Development Goals as Governance Innovation* edited by Norichika Kanie and Frank Biermann (2017).

The first book focuses primarily on the design of the SDGs and the politics behind them. It is comprehensive, and the critique is theoretically grounded. It discusses the side effects which the focus on goals can lead to, such as the unintended effects of a "quantification" of a policy, for example, that issues which are not easily measurable and/or selected for review are overlooked.

The second book identifies peculiar characteristics of the SDGs. The volume is at a high level of abstraction, taking a theoretical approach to the SDGs. It covers a range of problems which can occur in global policymaking, especially in the form of goal setting. It touches upon general problems of implementation, relating to drivers and barriers. One topic discussed is the importance of good governance as a foundation (driver) for sustainable development. Another topic is linkages (drivers) across policy areas and levels of the political system.

Empirical perspectives

The empirical type of books places the MDGs/SDGs in the context of ongoing change. They are often based on case studies of individual countries and bring out empirical data. However, the selection of countries is seldom motivated, which makes it difficult to see the more general picture of experiences and to draw conclusions on drivers and barriers which are applicable more generally. The texts are rather exploratory, identifying factors which deserve to be studied further. They provide some comments on the implementation of the global goals. In other words, they at least partly discuss the causal links from government policy to the outcomes (i.e. who did what, when and why?).

One example is *Globalization and Sustainable Development in Africa* edited by Bessie House-Soremekun and Toyin Falola (2011), which covers a broad range of topics, linking sustainable development in Africa to the broader debate about the pros and cons of globalization. It is an accessible background to the situation in Africa, especially in Nigeria, but there are few explicit links to the MDGs/SDGs.

Another empirical volume is *Implementation of the Millennium Development Goals. Progress and Challenge in Some African Countries* edited by Nicholas Awortwi and Herman Musahara (2015). This volume focuses on individual countries and issues (goals), which gives contextual information on relevant items, such as side effects and some drivers and barriers for the implementation of the MDGs. One chapter presents a conceptual framework for achieving one of the goals (p 167), which is an outline of the causal mechanisms involved. The scope of the volume is limited, but there is some depth in the coverage.

A third example is *The Millennium Development Goals: Challenges, Prospects and Opportunities* edited by Nana Poku and Jim Whitman (2014). It offers analysis of a limited selection of goals but on the other hand a larger selection of more general chapters. The first chapter outlines what was interesting about the MDGs as a strategy, arguing that it had to be general and adapted to each country. The second chapter identifies storylines in development strategies over time to show how the MDGs were different from previous strategies.

Another example is *Africa and the Millennium Development Goals. Progress, Problems and Prospects* edited by Charles Mutasa and Mark Paterson (2015). This volume focuses explicitly on the MDGs, presenting a chapter on each of the goals as well as on some general challenges. While each chapter has a focus, the variety across the chapters is great. Relevant for my study is, for example, a discussion about political and economic assumptions (p 24), development myths (p 26f) and the need for government departments to become learning organizations (p 120). The first chapters are critical to the MDGs as an outsider (Western) perspective on Africa.

A slightly different type of volume is *Development in Africa. Refocusing the Lens after the Millennium Development Goals* edited by George Kararach, Hany Besada and Timothy Shaw (2016). This is the most general of the empirical volumes, covering, for example, the problem of domestic resource mobilization – i.e. taxation (Chapter 7). The volume is aiming to draw general conclusions on the MDGs and to provide conclusions for the future, what became the SDGs. The analysis is on a general level, not going into specific cases.

Poverty reduction

Another type of empirical work focuses on the fundamental issues rather than the MDGs and SDGs. There is a great interest in the general topic of poverty reduction, which is a core issue, an underlying problem behind most of the goals. One example is the volume *The Last Mile in Ending Extreme Poverty* edited by Laurence Chandy, Hiroshi Kato and Homi Kharas (2015). This volume is designed as an overview of relevant research, mainly from economics. It belongs to a tradition of "practical economics", discussing what economics has to say about real-world problems. However, the level of technical language limits the audience to scholars.

Another example is *What Works for Africa's Poorest? Programmes and Policies for the Extreme Poor* edited by David Lawson, Lawrence Ado-Kofie and David Hulme (2017). This volume has another strategy, building on rich case studies which describe specific problems and situations. The questions are about poverty in general, not explicitly addressing the MDGs/SDGs. As with the previous edited volumes, there is no systematic selection of countries; hence, it is difficult to see the bigger picture.

The latter two volumes come close to policy learning in the sense that they talk about what has been done and what needs to be done next. In the case of poverty eradication, the proposed action is often something more radical to reach the remaining pockets of poverty. To put it differently, the books begin a discussion about what it takes to succeed in fulfilling these very ambitious goals. By doing so, they talk about drivers and barriers for development and change.

It is here that I see a need to connect the specific literature on the MDGs/SDGs with the more general literature on development in comparative politics and economics, dealing with issues such as governance and democracy promotion. This literature seldom connects with the SDGs, probably because the literature in comparative politics is mainly focussed on politics (democracy), rather than policy (social and other problems). Hence, the current book aims to fill a space by connecting several types of literature into a discussion about what is relevant to make the SDGs work.

As I will outline in the following chapters, national policy needs to make use of available research in a constructive way. My ambition to start filling the research gap on the strategies of implementation will begin with a discussion in the next chapter on how we can understand the implementation of the SDGs.

References

Andersen, Terry L. & Libecap, Gary D. 2014: *Environmental Markets: A Property Rights Approach*, Cambridge: Cambridge University Press.

Awortwi, Nicholas & Musahara, Herman (eds.) 2015: *Implementation of the Millennium Development Goals: Progress and Challenge in Some African Countries*, Addis Ababa: Organisation for Social Science Research in Eastern and Southern Africa.

Bernstein, Steven 2001: *The Compromise of Liberal Environmentalism*, New York: Columbia University Press.

Brown, Katrina 2016: *Resilience, Development and Global Change*, Abingdon: Routledge.

Chandy, Laurence, Kato, Hiroshi & Kharas, Homi (eds.) 2015: *The Last Mile in Ending Extreme Poverty*, Washington, DC: Brookings Institution Press.

Clapp, Jennifer & Dauvergne, Peter 2011: *Paths to a Green World: The Political Economy of the Global Environment*, Cambridge, MA: The MIT Press.

Cormier, Ben 2018: "Analyzing If and How International Organizations Contribute to the Sustainable Development Goals: Combining Power and Behavior", *Journal of Organizational Behavior*, vol. 39, pp. 545–558.

Fukuda-Parr, Sakiko 2017: *Millennium Development Goals: Ideas, Interests and Influence*, Abingdon: Routledge.

Funder, Mikkel, Mweemba, Carol & Nyambe, Imasiku 2018: "The Politics of Climate Change Adaptation in Development: Authority, Resource Control and State Intervention in Rural Zambia", *The Journal of Development Studies*, vol. 54, no. 1, pp. 30–46.

Hanson, Kobena T., Puplampu, Korbla P. & Shaw, Timothy M. (eds.) 2018: *From Millennium Development Goals to Sustainable Development Goals: Rethinking African Development*, Abingdon: Routledge.

Henley, David 2015: *Asia-Africa Development Divergence: A Question of Intent*, London: Zed Books.

Hill, Michael & Hupe, Peter 2002: *Implementing Public Policy: Governance in Theory and in Practice*, London: Sage.

House-Soremekun, Bessie & Falola, Toyin (eds.) 2011: *Globalization and Sustainable Development in Africa*, Rochester: University of Rochester Press.

Howlett, Michael & Ramesh, M. 1995: *Studying Public Policy: Policy Cycles and Policy Subsystems*, Oxford: Oxford University Press.

Kanie, Norichika & Biermann, Frank (eds.) 2017: *Governing through Goals: Sustainable Development Goals as Governance Innovation*, Cambridge, MA: The MIT Press.

Kararach, George, Besada, Hany & Shaw, Timothy M. (eds.) 2016: *Development in Africa: Refocusing the Lens after the Millennium Development Goals*, Bristol: Policy Press.

Landman, Todd 2013: *Issues and Methods in Comparative Politics: An Introduction*, Third edition, Abingdon: Routledge.

Lawson, David, Ado-Kofie, Lawrence & Hulme, David (eds.) 2017: *What Works for Africa's Poorest? Programmes and Policies for the Extreme Poor*, Bourton on Dunsmore: Practical Action Publishing.

Miller, Raymond C. 2018: *International Political Economy: Contrasting World Views*, Second edition, Abingdon: Routledge.

Mutasa, Charles & Paterson, Mark (eds.) 2015: *Africa and the Millennium Development Goals: Progress, Problems and Prospects*, London: Rowman & Littlefield.

Newell, Peter 2012: *Globalization and the Environment: Capitalism, Ecology and Power*, Cambridge: Polity Press.

Poku, Nana & Whitman, Jim (eds.) 2014: *The Millennium Development Goals: Challenges, Prospects and Opportunities*, Abingdon: Routledge (previously published as *Third World Quarterly*, 32:1).

Rodrik, Dani 2011: *The Globalization Paradox: Democracy and the Future of the World Economy*, New York: W.W. Norton & Company.

Sachs, Jeffrey D. 2015: *The Age of Sustainable Development*, New York: Columbia University Press.

Vedung, Evert 1982: *Political Reasoning*, London: Sage.

3 Refocus from the goals to learning over time

The previous chapter defined this study as forward-looking, focussing on the design and implementation of the SDGs. This chapter will elaborate on what a focus on design and implementation entails. It is intended to make explicit the theoretical perspective from which I study the SDGs. In other words, this is the analytical framework of the study. It will point out how the study is conducted and why I have chosen this perspective rather than any other perspective.

An analytical framework is mainly a set of assumptions about what is important to study, how to do it and how to draw conclusions, especially if explanatory theories are used. Theories in political science often make assumptions about motives and intentions, which is a basis for drawing conclusions about the behaviour of actors. I will only assume some rationality and, most of all, that policies have a purpose and an intended instrumental design (Rose 2005; Dodds 2013).

My perspective is that a study of the design and implementation of the SDGs needs to elaborate on the intended causal mechanisms of the policy in order to understand how it is meant to work and (subsequently) compare it to what happens in reality. Especially when the study is done at the start of the implementation ("ex ante"), the focus should be on how it is intended to work. The expected outcomes of such an approach is a better image of what the policy is, together with suggestions for improvement and issues to follow more closely later on.

Hence, my focus is on drivers and barriers (causal mechanisms) in order to make predictions on how the SDGs will work. Most of my discussion is about tensions within the SDGs and in relation to contextual factors. The following chapters will scrutinize the SDGs in line with what is said here (Chapter 4), compare the SDGs to alternative designs (Chapter 5) and discuss the key issue of creating an organization to fulfil the implementation of the SDGs (Chapter 6). Chapter 7 (by Tobias Ogweno) will look at the national implementation of the SDGs.

Policy by goals

Most analyses of the SDGs focus on the fact that they are goals, which is assumed to have certain consequences for the policies (Fukuda-Parr 2017; Kanie & Biermann 2017). I will propose an alternative perspective to avoid some of the problems related to goals, a softer and less demanding approach. This is the idea of a learning policy, where the focus is on adjustments over time, to improve the implementation of the policies. This learning will be meaningful if there is an idea of how the policy is intended to work so that the intention (the strategy) can be improved in light of experience.

The role of goals

The SDGs indicate a future state of affairs which is desirable (and some of the means to reach these goals, to be discussed in the next chapter). This vision (goal) is the main component of the policy. The explicit use of goals is in contrast with other policies which focus on the specific activities which need to be carried out. It is also in contrast with the academic debate, which is to a large extent about understanding drivers and barriers, often with the help of one or several general (theoretical) perspectives on poverty and the environment.

The goals not only simplify the bargaining situation but are also thought by proponents of goal-based policies to simplify the implementation. The goals select a focus for the implementation, which is thought to make policies more efficient. By doing so, the goals seem to make it easier to measure achievements and hold actors and organizations accountable, perhaps also to redesign policies if they don't work. One of the purposes of this kind of design is to put stronger force on the implementing organization (Radin 2006). This assumes that it is relatively easy to identify appropriate indicators to measure complex phenomena like development. It also assumes that organizations have the power to influence the outcomes with their instruments ("outputs"). The problem of fragmented and/or overlapping responsibilities among organizations is ignored (e.g. Torfing & Triantafillou 2011).

Another aspect is that goals make it easier for politicians to explain what they are doing, though there is also a risk is that they talk too much about the future state of affairs and not how to reach it. In other words, goals make governments look good by talking about ambitions while they avoid difficult questions about implementation or even an analysis of what the key problems (drivers and barriers) are. The goals could furthermore empower bureaucracies and consultants to design the specific policies – i.e. the means to reach the ends. Goals also have unintended side effects, such as changing the way we think about development (Fukuda-Parr 2017).

The targets of the SDGs can be seen as intermediary goals to achieve the main goals, though there is no explicit discussion about their relation to the main goals. It is up to studies like this one to discuss whether the targets will lead to the fulfilment of the main goals and how they relate to other goals more than the ones they are listed under.

The focus on goals can open up experimentation by national governments and comparisons of what works in terms of promoting sustainable development. The global policy can evolve into something of a scientific testing of alternative strategies. However, the complexity of the issues could also lead to ritual implementation of activities which are popular in the global community. Copying what others do may be the least demanding strategy.

The difference between ritual implementation and a reasonable focus on what works has to do with the mechanisms in place for analyzing what is done, if there can be serious learning about what works and a serious interest in modifying the strategies in line with the experience.

New Public Management

The focus on goals is an example of results-based management, which has been applied by national governments for a long time. There is a strong debate on unintended side effects of this way of designing policies, often referred to as New Public Management (NPM) (e.g. Radin 2006; Hood 1994, 1998). In more general terms, NPM is based on economic theories which emphasize the importance of incentives ("neoliberalism"). The problems arise when it is difficult to design the incentives or where they clash with other norms.

There are many versions of the debate, but it is generally about the pros and cons of focussing on the intended outcomes of a policy, rather than on the means to reach them. Some see it as an advantage that governments don't specify how the goals should be reached, while others see it as wishful thinking since the goals don't have to be realistic. The element of learning over time is often forgotten by governments.

In theory, governments – in this case the UN – should make a distinction between ends and means, and only regulate the ends while leaving the choice of means open for the implementing organization (the national governments). A common problem with NPM is that governments also regulate many of the means to reach the goals. This is especially the case when the means are controversial. It may be easier to agree on a goal like good health than to agree on the means to reach it – for example, the role of public and private providers. In reality, governments may regulate such contested issues, which creates a mixed policy of ends and means.

The SDGs actually consist of a mix of ends and means, as I will discuss in the next chapter. The social and environmental goals are primarily about ends, while the economic and cross-cutting (governance) goals give much room for means to reach other goals.

The use of indicators has further effects on national policies. The issues which get measured will become more important than other issues. National governments often tie funding to activities, though they are only a means to reach the goals. This could lead to goal displacement, where the implementing organization focuses on what gets measured and/or funded and forgets about other dimensions of the general goal (ibid.). It is difficult to tie the funding to the fulfilment of goals since there are often many other factors which have an impact on the outcome.

In the SDGs, there are risks with the selection of goals, targets and indicators. National governments will probably focus on the issues they are asked to report on. If they forget about the context, the wider picture, they may cause unintended effects. There is also a problem with compliance related to the legitimacy of the SDGs. The pressure to comply is stronger if there is loyalty among the governments and if citizens perceive the SDGs as respectable (legitimate).

Another interpretation: a learning policy

Goals can be interpreted as part of another tradition, partly compatible with NPM. This is the more pragmatic tradition of seeing politics as problem-solving over time (e.g. Ansell 2011). Here, the key is not the goal itself, but rather the follow-up and analysis to draw conclusions about adjustments. In this perspective, the goals are the starting point and the end of the policy process, but the emphasis shifts to what is done in between.

Learning is a concept from organizational theory. It is about adjusting beliefs to change future action (Cooke & Morgan 1998; Friedmann 1987; Schön 1983). The key is a new understanding of how to reach something (a causal relationship). It may turn out that the intended mechanism didn't work because some other factor was missing. That would be an example of gaining a more complete understanding of the causal mechanisms involved. Such learning is in contrast with revising policies without an improved understanding of the situation, for example, if it is based on doing what others do – i.e. mimicking what appears to be interesting or successful.

The learning is mainly over the instruments applied to reach the goals. The choice of instruments can be improved over time by looking at the effects of the instruments and other key variables which affect the outcomes. The instruments need to be understood in context and other relevant variables need to be identified. To evaluate a policy is partly to understand the causality of how it works (e.g. Owen & Rogers 1999; Pawson & Tilley 1997).

This learning exercise may be overwhelming for individual governments, but the SDGs can open up a fruitful discussion and comparison across countries of what works. The governments can make use of any relevant experiences, as well as theories from research.

Rationalism vs realism

This perspective is based on a rationalistic understanding of governments. It is assumed that governments want to take rational action and that they are able to do so. This stands in contrast to the more realistic understanding of politics as a process of "muddling through", dealing with issues from day to day (Lindblom 1959). In reality, politicians may find it difficult to deal with all expected issues. It is more efficient to avoid complexity and do things which signal an interest in the issues which citizens are concerned with, even if politicians lack the means to solve the specific problems (Brunsson 2003).

Implementation of policies is based on means-ends thinking (rational instrumentalism), while politics is also about taking stands, showing sympathy and surviving to the next election. Global policies and national policies may be adopted to show concern rather than actually trying to solve problems.

There are many risks with the SDGs since they deal with very complex matters. Hence, there is a need to help politicians take the ambitious road rather than ritual implementation of the SDGs.

Another criticism of the rationalistic approach is that governments overestimate their capacity to govern. One explanation for failure is the difficulty of assembling all the relevant information at the centre. The UN has taken a softer approach with the SDGs, avoiding instructing national governments on what to do. This role may be played by donor organizations and others.

I am assuming that governments want to think rationally about a plan to implement the SDGs. In other words, I assume they think about the situation in their country, what needs to be improved and how. They need to think about barriers – what keeps their country from reaching the goals – and drivers for change – what instruments can change the situation. This is an analytical abstraction. In reality, government ministers may find it the most rational not to act according to a plan.

Critics argue that governments and other organizations pretend to be rational planners because that is what outsiders expect and reward leaders for. In the area of sustainable development, there is on the one hand a need for rethinking many policies which clash with the SDGs – i.e. taking bold and rational action. On the other hand, the complexity of the issues makes it difficult to design smart plans and to avoid criticism by the public and by international organizations and others. The complexities could stimulate action as well as "muddling through", waiting to see, for example, what other countries do.

By assuming rationality, I treat the issue of sustainable development in a technocratic fashion. This is a simplified but nevertheless necessary starting point of the analysis. It is more of an academic analysis of the situation than an analysis of national governments and how they deal with the situation. At this point, I will only note the difference between the academic perspective and the perspective of the national governments, even though the latter may be highly relevant in explaining, for example, why governments do things other than what academic research suggests.

A learning subject

Learning must be by someone. The key actors in the implementation of the SDGs are the national governments since they are formally in charge of designing and implementing policies. In reality, it can be very difficult to learn within an elite group of politicians or across ministerial boundaries. In a democracy, one can also think of political parties as learning subjects since they are the organizations which propose future policies to the electorate.

As I mentioned, this learning activity may in fact be a combination of looking at what others do and some limited thinking on the need for reversing policies. Rationality in the sense of thinking through ends and means in light of experience may be too demanding, an unrealistic assumption if we want to understand real-life politics. "Muddling through" may be the best approach to complex matters.

Another part is to ask consultants to investigate and propose new policies, which creates a risk that the produced documents will reflect the knowledge of the consultants rather than their clients, the national governments. Reports can give a brushed-up image of what goes on, by using language which is liked by outsiders, without producing learning in the national government.

Learning can also take place at the global stage, by the UN as a coordinating body for information on the implementation of the SDGs. The High-Level Political Forum (HLPF) is an organ to oversee and draw lessons from the process. The United Nations Development Program (UNDP) is involved in giving advice to national governments on what to do. Again, there are temptations to stay on the surface rather than digging deeper into what really goes on in individual countries and what the effects are.

A technical difficulty in all of this is to establish the causal effects, rather than just describing what has happened. A description of events could be a first step in trying to understand how the policies produced outcomes. The latter involves thinking about alternative causes and alternative effects (side effects). In research, this is often done through comparison with a situation where the policy in question has not been followed – i.e. a comparison group (Landman 2013).

The importance of a programme theory, a strategy

Learning must be informed. Data on a certain situation is seldom of interest without an interpretation. Likewise, conclusions are difficult to draw without an interpretation. One type of interpretation is to have an idea, a hypothesis, about what should happen. Such hypotheses can be improved over time to take into account theories and experiences. Unexpected data might lead to a revision of the understanding of the situation and a change of instruments. This would be an example of learning.

The key is to think about drivers and barriers, and to revise this thinking over time. Such thinking is called a theory of change, a programme theory or a logical framework ("logframe"; Owen & Rogers 1999). It spells out the steps that need to be executed to reach the goal. One technique is to develop a roadmap to spell out which steps are necessary to reach the goal in time. An example of a theory of change is a causal chain, where the first action (X) brings about (Y), which in turn brings about (Z) (Figure 3.1).

A preliminary theory of change (a hypothesis) can be constructed out of an intuitive understanding of a problem and what needs to be done. The more interesting analytical work is to check if this intuitive understanding is correct. There may be side effects which are relevant to take into consideration – i.e. that (X) also leads to (P), which might be something undesirable in relation to (Z), a negative side effect.

Hidden assumptions

There may also be overlooked assumptions which need to be spelled out. An example is if (X) can only produce (Y) under certain circumstances – i.e. another factor (Q) is necessary too. It is (X) and (Q) which together bring about (Y). This is probably one of the most useful uses of spelling out a theory of change: to show hidden assumptions in the original hypothesis. Such assumptions are steps that should be part of the outline of the causal chain.

| X | → | Y | → | Z |

Figure 3.1 A causal chain of events

Several authors have looked for a hidden plan in the SDGs – for example, by rearranging them to find common elements or an implied order of priority (e.g. Nilsson 2017). This has some similarities with the ambition to find a plan in the concept of sustainable development, which showed that there were numerous alternative interpretations (Kates, Parris & Leiserowitz 2005). There is also a growing literature on what it takes to reach the SDGs, how national governments design strategies, on their own or together with international organizations. This literature can be read as suggestions for an elaborate programme theory ("theory of change"), spelling out the missing elements of national strategies of implementation. I will do something similar in the next chapter.

The SDGs are highly complex. There is growing research on how to prioritize and how to make use of links across the goals. These are examples of questions that need to be asked in order to understand what needs to be done about very complex issues, such as increasing economic growth and reducing the impact on the environment at the same time. The SDGs are about highly contested political issues which need to be discussed. Very complex relationships can be analyzed with the help of theories from the social sciences.

A theory of change for sustainable development

The SDGs need to be arranged in a hierarchical fashion, with ultimate goals at the top and intermediary goals at a lower level. Poverty reduction and environmental protection are the ultimate goals of the SDGs (see below). The specific goals related to each can be seen as intermediary goals. The cross-cutting goals are prerequisites, hidden assumptions to reach the other goals.

The drivers and barriers are the keys to working out rational steps to achieve the goals. Some of the goals are very specific, such as a providing food, water, sanitation, health care and education. A rational plan would discuss ways to combat diseases, for example. Immediate solutions might be to distribute preventive instruments, such as mosquito nets, to stop the spreading of malaria. Long-term solutions would include the training of doctors and nurses, setting up hospitals and so on. This, in turn, would make it necessary to raise the funds and, ultimately, to have an organization which is able and trusted to raise the funds and deliver the services. These few examples indicate the wide range of issues involved and the need to think about costs and effects.

Many of these things are already done by national governments and others. The question is rather how it can be improved and pursued on a larger scale within the nexus of sustainable development (economic, social and environmental ambitions). Improved health is a means (a driver) for economic development, but economic resources are at the same time a pre-requisite (a driver) for making social investments. Some actions may be harmful (barriers) to the environmental goals, while other are neutral or have positive effects (drivers) on the environmental goals.

Other goals are more complex, such as "providing decent work for all" (no. 8) or "combat climate change" (no. 13). These goals need more unpacking in terms of implied meanings, values, causal mechanisms, tentative options, etc. Some of the goals are instrumental (drivers) to achieve the more specific goals, especially the ones which deal with the production and conservation of resources (growth and protection of the environment). A rational government would need to deal with these general questions of sustainable development.

The key question for a rational plan is to work out how the drivers and barriers relate to each other, especially if there are underlying relationships which have an impact on many issues. One such issue is the role of economic growth, as a means to make more resources available, but also as a potential threat to the environment and traditional lifestyles, etc.

A "theory of a change" is a map of arrows, starting from the goals and working backwards to identify drivers and barriers. For example, what is needed to achieve the goals of improved health, decent work or a limit on climate change? If the necessary actions conflict with each other, how should they be prioritized? Are there balanced strategies which would achieve a reasonable set of goals? An ideal process of rational planning would aim to spell out these causal chains, including any synergies and conflicts.

It would make sense to use drivers which are already present, such as interaction of sellers and buyers in a food market. Buyers and sellers, investors and savers make decisions all the time based on current prices and expectations about the future. It would be wise to build on market coordination and only aim to change it where it is considered necessary. For example, a central idea behind the MDGs was that the poorest people mainly need a little help to get the process of economic development going (UN Millennium Project 2005). The thinking was that they are trapped in a situation where they are literally outside the workings of the economy and where they need to do things which are bad for the environment, like burning manure. With some help from the outside, the poorest people would start improving their own situation. I will come back to the role of the market as a coordinating device and how government planning relates to it.

Methods and material

In the next chapter, I will go through all the goals and their targets to see if there is an implied strategy (theory of change) in the document. I will primarily look for evidence of drivers and barriers in the text. It will be a close reading of the SDGs – an effort to reflect on and spell out more clearly what the document says.

Chapter 5 will construct alternative perspectives to show more clearly what the options are and where the SDGs side. I will use a method of contrasting ideal types from IPE to show pure positions as a map of a policy space for policies for sustainable development (Miller 2018). I will use this to classify the positions as liberal, Marxist or nationalist (statist). Here the interpretations are reconstructed abstract generalizations, even simplifications of positions in the debate intended to show the great variety of possible drivers and barriers.

Chapter 6 is more empirical, digging deeper into the question of how important good governance is as a prerequisite for the other goals. I will do this by looking at the academic debate on the issue, starting from a position which argues that good governance isn't as important as the World Bank and others believe. This analysis will discuss the thinking and the evidence presented in the debate, as well as the clash of values involved. It is intended to show the complexity of the issues, especially when asking where developing countries should start their policies for sustainable development.

The three chapters together form an overview of some of the most important issues to deal with in an analysis, beforehand ("ex ante"), of the implementation of the SDGs. Tobias Ogweno will deal with some questions about what goes on nationally, such as the contents of the national plans and the organizations set up to implement them. Further questions for future investigation have to do with what happens on the ground when the SDGs are implemented. An example of such a study is the previously mentioned analysis of the impact on rural Zambia (Funder, Mweemba & Nyambe 2018).

References

Ansell, Christopher K. 2011: *Pragmatist Democracy: Evolutionary Learning as Public Philosophy*, Oxford: Oxford University Press.

Brunsson, Nils 2003: *The Organization of Hypocrisy: Talk, Decisions and Actions in Organizations*, Stockholm: Liber.

Cooke, Philip & Morgan, Kevin 1998: *The Associational Economy: Firms, Regions, and Innovation*, Oxford: Oxford University Press.

Dodds, Anneliese 2013: *Comparative Public Policy*, Houndmills: Palgrave Macmillan.

Friedmann, John 1987: *Planning in the Public Domain:. From Knowledge to Action*, Princeton: Princeton University Press.

Fukuda-Parr, Sakiko 2017: *Millennium Development Goals: Ideas, Interests and Influence*, Abingdon: Routledge.

Funder, Mikkel, Mweemba, Carol & Nyambe, Imasiku 2018: "The Politics of Climate Change Adaptation in Development: Authority, Resource Control and

State Intervention in Rural Zambia", *The Journal of Development Studies*, vol. 54, no. 1, pp. 30–46.

Hood, Christopher 1994: *Explaining Economic Policy Reversals*, Buckingham: Open University Press.

Hood, Christopher 1998: *The Art of the State: Culture, Rhetoric, and Public Management*, Oxford: Oxford University Press.

Kanie, Norichika & Biermann, Frank (eds.) 2017: *Governing through Goals: Sustainable Development Goals as Governance Innovation*, Cambridge, MA: The MIT Press.

Kates, Robert W., Parris, Thomas M. & Leiserowitz, Anthony A. 2005: "What Is Sustainable Development? Goals, Indicators, Values and Practice", *Environment: Science and Policy for Sustainable Development*, vol. 47, no. 3, pp. 8–21.

Landman, Todd 2013: *Issues and Methods in Comparative Politics: An Introduction*, Third edition, Abingdon: Routledge.

Lindblom, Charles 1959: "The Science of 'Muddling Through'", *Public Administration Review*, vol. 19, no. 2, pp. 79–88.

Miller, Raymond C. 2018: *International Political Economy: Contrasting World Views*, Second edition, Abingdon: Routledge.

Nilsson, Måns 2017: *Important Interactions among the Sustainable Development Goals under Review at the High-Level Political Forum 2017*, Working paper 2017–06, Stockholm: Stockholm Environment Institute.

Owen, John M. & Rogers, Patricia J. 1999: *Program Evaluation. Forms and Approaches*, London: Sage.

Pawson, Ray & Tilley, Nick 1997: *Realistic Evaluation*, London: Sage.

Radin, Baryl A. 2006: *Challenging the Performance Movement: Accountability, Complexity and Democratic Values*, Washington, DC: Georgetown University Press.

Rose, Richard 2005: *Learning from Comparative Public Policy: A Practical Guide*, Abingdon: Routledge.

Schön, Donald A. 1983: *The Reflective Practitioner: How Professionals Think in Action*, Farnham: Ashgate.

Torfing, Jacob & Triantafillou, Peter (eds.) 2011: *Interactive Policy Making, Metagovernance and Democracy*, Colchester: ECPR Press.

United Nations Millennium Project 2005: *Investing in Development: A Practical Plan to Achieve the Millennium Development Goals*, New York: Earthscan.

4 An inherent strategy in the goals

In the previous chapter, I introduced the idea of designing a plan (a so-called programme theory or a theory of change) for the national implementation of the SDGs. I also proposed that it should be tentative, with hypotheses about what governments should do and with a focus on piecemeal adjustments by learning from experience. In this chapter, I will discuss if there is such a plan inherent in the SDGs (UN 2015a).

The chapter will go through the goals in detail, looking at the specific targets and some of the indicators, if they add something which isn't obvious in the text of the target. The general question is, what is the message of the goals in terms of ends and means? Are the ends to be achieved clearly stated? When it comes to the means, are there any hints of means to achieve the ends? Are there, in other words, factors which imply a causal relationship and, hence, a strategy? What does such an implied strategy look like?

I will divide the goals into four groups, as in Chapter 1, to make it easier to see the bigger picture. I will first discuss the economic dimension, the social dimension and the environmental dimension. Lastly, I will discuss the cross-cutting issues ("the governance dimension"). I will point out links between the goals, and I will comment on them as a package at the end of the chapter. How can we understand the total set of goals? How are they interlinked?

The economic dimension

The four goals which are primarily focussed on the economy are the goals 1, 8, 9 and 10. These goals talk about the means to reduce poverty – i.e. to increase wealth in poor countries. The next group, which deals with the social dimension, goes into more detail about what it means to live in poverty and what some of the specific problems are.

Goal 1: no poverty

The first goal states one of the general aims of the SDGs: to "end poverty in all its forms everywhere". The scope of the goal is very demanding, and the targets specify it further. There are five targets with numbers, indicating a causal relationship with the goal, and two targets with letters, indicating "means of implementation". The first targets state the general ambition. Target 1 defines extreme poverty as living on less than $1.25 a day (now adjusted to $1.90). The second target adds an ambition to halve the remaining poverty as defined nationally.

Several targets bring out links to other goals. Targets 1.3 and 1.5 talk about social protection systems (the social dimension) and resilience in the event of disasters of people who are poor and in vulnerable situations (the environmental dimension).

The most specific statement on how to overcome poverty is in target 1.4, which mentions some possible drivers and barriers for poverty reduction (i.e. for economic development). It begins by stating that there should be "equal rights to economic resources as well as access to basic services". This could mean that governments should provide some services (in the form of rights for the citizens), but it could also mean that governments should stop discriminating against the poor. The text, so far, is rather vague. Indicator 1.4.1 talks about the proportion of the population living in households with access to basic services, not further specified.

The rest of the sentence in target 1.4 mentions factors which are important for economic development: "Ownership and control over land and other forms of property, inheritance, natural resources, appropriate new technology and financial services, including microfinance". At least the first factors seem to indicate that the poor are discriminated against by not owning the means for economic development. Indicator 1.4.2 supports this interpretation by asking about the "proportion of total adult population with secure tenure rights to land, with legally recognized documentation and who perceive their rights to land as secure". The basic message, then, is that the poor should have legal rights to their living conditions. This can be interpreted as a statement on the fundamental conditions for economic development, for farming and for running a business (e.g. de Soto 2000).

Target 1.A on "the means of implementation" talks about the mobilization of resources and mentions foreign aid, while domestic taxes are not mentioned in this context. The indicators for this target focus on the profile of government spending. They talk about the "proportion of resources allocated by the government directly to poverty reduction programmes" and the "proportion of total government spending on essential services (education, health and social protection)". Target 1.B and its indicator go further by focussing on spending for "women, the poor and vulnerable groups". These targets are interesting in relation to the targets on the social dimension (below), where the funding of social programmes is not discussed.

To summarize, the main messages of goal 1 is that economic growth is important and that discrimination against the poor must end. This is in line with the debate on the empowerment of the poor, where it is argued that the poor are often denied the legal means for improving their lives (de Soto 2000).

A final comment on goal 1 concerns what it doesn't say. It is noteworthy that the text avoids alternative ways of formulating the problem of poverty, such as vulnerability, loss of power, autonomy or dignity. The text mainly uses more neutral terms.

Goal 8: decent work and economic growth

Goal 8 continues with the issue of economic growth. The goal is to "promote sustained, inclusive, and sustainable economic growth, full and productive employment, and decent work for all". The targets for goal 8 add interesting specifications, such as a minimum growth of gross domestic product per capita of 7 per cent in the least developed countries (target 8.1). This is strong support for economic growth, perhaps unexpectedly strong in a document on sustainable development.

The other targets for goal 8 mention specific concepts related to the economy, such as productivity and entrepreneurship. These are factors which have an impact on economic growth, which is to say that they talk about the means for achieving economic growth and imply a strategy for achieving it.

Target 8.2 elaborates on some specific ways to increase productivity "through diversification, technological upgrading and innovation, including through a focus on high-value added and labour-intensive sectors". The role of small firms is mentioned in target 8.3, which elaborates on the policies to achieve economic development: "Support productive activities, decent job creation, entrepreneurship, creativity and innovation, and encourage the formalization and growth of micro-, small-and medium-sized enterprises, including through access to financial services".

Target 8.4 balances the support for economic growth by stating two important principles: that industrialization shouldn't increase the burden on the environment and that the industrialized countries should show the way. The target is to "improve progressively, through 2030, global resource efficiency in consumption and production and endeavour to decouple economic growth from environmental degradation . . . with developed countries taking the lead".

Targets 8.5 through 8.8 relate to employment issues, such as decent work, including putting an end to child labour, increasing occupational safety and respecting collective bargaining. Sustainable tourism is mentioned in a specific target (target 8.9) and so is access to financial services (target 8.10).

The final two targets indicate important "means of implementation" in the form of trade ("Aid for Trade"), a global strategy for youth employment and for the Global Jobs Pact by the ILO (ILO 2009). The reference to trade suggests that poor countries can develop more quickly if they are part of global value chains, producing goods for the global market, which is a point that comes back in other goals (e.g. Glenn 2007).

To summarize, goal 8 takes a strong position in favour of economic development and spells out some of the factors which are important for the national economy. Productivity, entrepreneurship and employment are three key factors. The goal also points out restrictions in terms of reducing the burden on the environment. Whether this can in fact be done, or if it is a case of wishful thinking, is one of the big debates regarding the SDGs.

Goal 9: industry, innovation and infrastructure

The issue of goal 9 is to "build resilient infrastructure, promote inclusive and sustainable industrialization, and foster innovation". This goal is strongly interrelated with the previous goal since industrialization is an important means to create employment and economic growth. This is a controversial position among environmentalists, and it is here that concepts such as "sustained, inclusive, and sustainable" are often mentioned.

The targets mention infrastructure (target 9.1), which is important for industry, as well as for farmers and others for the transport of resources and products. Industrialization and industrial employment are mentioned in target 9.2. Target 9.3 mentions financial services (affordable credit) and "integration into value chains and markets". This is explicit support for the idea that global trade and economic integration are good for developing countries.

Target 9.4 adds the concern for the environment "with increased resource-use efficiency and greater adoption of clean and environmentally sound technologies and industrial processes, with all countries taking action in accordance with their respective capabilities".

Target 9.5 addresses technological upgrading and innovation, which are key activities within firms to adjust to changing circumstances and to keep up the competitiveness, as well as reducing the impact on the environment. The text is interesting in the way it talks about a goal (upgrading) and the means to do it (the funding and the number of people employed in scientific research). The link from research to the technological upgrading of firms is only mentioned to "promote innovation", while the literature on innovation emphasizes how difficult this is (Lerner 2009). In other words, there is some wishful thinking that increased spending will automatically lead to technological upgrading and increased competitiveness.

The targets on the means of implementation talk about "enhanced financial, technological and technical support" (target 9.A), as well as "domestic technology development, research and innovation in developing countries" (target 9.B). The latter points to other important factors besides research and development (R&D) – namely, "inter alia, industrial diversification and value addition to commodities" (ibid.). Target 9.C talks about "access to information and communications technology and . . . universal and affordable access to the Internet in least developed countries by 2020".

To summarize, goal 9 is very similar to goal 8 in that it outlines important factors for economic growth and balances this with concerns for the environment. The key factors are, in the first case, industrialization and trade, while technological development is important for economic growth as well as for environmental protection.

Goal 10: reduced inequalities

The SDGs have a goal (goal 10) which addresses inequality, with an ambition to "reduce inequality within and among countries". This is a concept open to several interpretations. The problem of absolute poverty – living under $1.90 a day – is dealt with in goal 1, while this goal on inequality is rather about relative poverty (and relative wealth) – i.e. the difference between the rich and the poor.

The issue of relative poverty is a traditional divider between the political left and right, where the left argues that inequality is undesirable in itself and leads to bad consequences, such as tensions in society (Piketty 2014). The right argues the opposite, that there are other values which are more important (such as individual freedom) and that inequalities, especially when they are rewards for individual achievements, lead to good consequences as part of a wealth-creating system (e.g. Rawls 1971). A fundamental disagreement is to what extent differences which are due to individual choices and lifestyles are legitimate or should be corrected through welfare systems (Kymlicka 2002).

Leaving the philosophical interpretations aside, the term "reduce" in goal 10 suggests that all governments think that differences are now too big. The terms "within and among countries" suggest that it is not just about the relations between the global north and south. It also applies to the domestic situation in countries that include China and India, which have large amounts of rich as well as poor people. It could even suggest that poverty reduction in the emerging economies should be more of a domestic concern than a global concern (Sumner 2016).

The targets avoid the philosophical issues and settle for an ambition that the poorest 40 per cent of the population should have a higher rate of growth than the national average (target 10.1). Target 10.3 states the more general principle, to "ensure equal opportunity and reduce inequalities of outcome". Target 10.4 talks about adopting policies to progressively achieve greater equality.

Other targets of this goal interestingly highlight barriers (discrimination) to be removed rather than actions to reduce differences, which is attractive also to those who think inequality isn't necessarily a problem. Target 10.2 talks about "the social, economic and political inclusion of all". Target 10.3 adds "eliminating discriminatory laws, policies and practices", which connects with similar issues in goal 1. Indicator 10.3.1 specifies this further by making reference to international human rights law.

A special case is the problem with global migration addressed in target 10.7 by saying that the global society should "facilitate orderly, safe, regular and responsible migration and mobility of people, including through the implementation of planned and well-managed migration policies".

There are furthermore targets relating to global power under this goal – e.g. on "enhanced representation and voice for developing countries in global international economic and financial institutions" (10.6), and on "special and differential treatment for developing countries" (10.A). Other targets focus on "better regulation . . . of global financial markets" (10.5) and lower transaction costs for migrants sending money home though remittances (10.C).

Only one target deals with the relationship between rich and poor countries in the form of aid: "Encourage official development assistance and financial flows, including foreign direct investment, to States where the need is greatest" (target 10.B).

To summarize, the goal talks about equal opportunities and reduced inequalities of outcomes. The main problems (causes) it points to have to do with discrimination against the poor. There is very little on the relationship between the global north and south, except for some detailed targets about power in international organizations.

Summary

To summarize what the SDGs say on the economic dimension, the four goals emphasize economic growth and add some qualifications, such as a discussion about discrimination as well as inequality. They also add some specifications of the means to achieve growth, such as industrialization, trade, increasing productivity and technological upgrading.

The four goals on economic development can be summarized in six points:

- There is strong support for economic growth and industrialization in a sustainable fashion
- Technological upgrading is important
- The concept of equality is mainly focussed on improving conditions for the lower 40 per cent (not on explicitly reducing the rich)
- Government discrimination against the poor must end
- Government spending must focus on the poor
- Poor countries must be treated more fairly in international organizations

The drivers and barriers hinted at come mainly from economic thinking on what drives growth, including the problem of discrimination against the poor.

The social dimension

There are seven goals which address social issues, the goals 2–7 and 11. Goal 2 has some overlap with the goals on the economic dimension, while goal 11 has some overlap with the environmental dimension. All of them are further specifications of extreme poverty (goal 1) in the sense that to be extremely poor means that you are very likely to suffer from problems such as hunger and diseases.

These goals can be described as welfare issues, something where each country makes decisions on services to be provided, either through public or private means. The immediate need in poor countries is to solve urgent problems; however, increased economic growth will provide the means for a long-term perspective on these issues. Most countries already have some kind of welfare systems (Rudra 2008).

These goals are in general much more focussed on stating the aims of the goals, than the goals on the economic dimension are. Most of the targets and indicators specify dimensions of each goal and how they should be measured. There is almost nothing that indicates how the goals can be fulfilled. This means that they are consistent with the idea of specifying ends, not means. The major exception is goal 2, which indicates ways to improve agriculture (which is an important part of the economy) and sounds much like the goals on economic development in the way it mentions causal factors.

The means to achieve the social goals are at best mentioned in the economic goals (noted earlier) or implied in the final goals on cross-cutting issues, which I will come back to.

Goal 2: zero hunger

Goal 2 is to "end hunger, achieve food security and improved nutrition, and promote sustainable agriculture". The targets for this goal differ from the other goals in this group by including factors which can be seen as remedies (implied causes) of hunger (Kamau, Chasek & O'Connor 2018:164). The factors are, for example, agricultural productivity; "double the agricultural productivity and incomes of small-scale food producers" (target 2.3). Target 2.3 further mentions access to land and other inputs, as well as to markets and opportunities for value addition, all of which are means to bring about the goal.

Other targets connect to the environmental goals by means of the concepts "sustainable food production systems" and "resilient agricultural practices", which are believed to lead to increased "productivity and production, that help maintain ecosystems, that strengthen capacity for adaptation to climate change, extreme weather, drought, flooding and other disasters and that progressively improve land and soil quality" (target 2.4). Target 2.5 mentions genetic diversity of seeds, plants and animals.

The "means of implementation" lists three issues: technological development (target 2.A), the removal of trade restrictions (target 2.B) and "measures to ensure the proper functioning of food commodity markets and their derivatives" (target 2.C).

To summarize, goal 2 focuses on hunger, which is closely related to the issue of extreme poverty (goal 1). It points to factors to increase food production within environmental limits. It is in this sense parallel to goals 8 and 9.

Goal 3: good health and well-being

The aim of goal 3 is to "ensure healthy lives and promote well-being for all at all ages", which is, of course, an urgent need among the poorest people. The targets elaborate on a number of specific issues, such as maternal mortality (deaths related to giving birth; target 3.1) and child mortality (deaths from birth until five years of age; target 3.2). Target 3.3 mentions diseases, such as AIDS and malaria, while target 3.4 mentions non-communicable diseases. Target 3.8 is one of the most general, stating the goal of universal health care.

A number of lifestyle risks are addressed. Target 3.5 addresses substance abuse, such as drugs and alcohol, while target 3.A addresses the use of tobacco. Target 3.6 talks about traffic deaths. Target 3.9 relates to the environmental goals in terms of "hazardous chemicals and air, water and soil pollution and contamination".

One of the most controversial issues during the negotiations on the SDGs was universal access to sexual and reproductive health-care services, including family planning (Kamau, Chasek & O'Connor 2018:166). This is mentioned in target 3.7 as well as target 5.6. The indicators will measure the use of "modern methods" (indicator 3.7.1) as well as the adolescent birth rate – i.e. girls/women 10–19 years old giving birth (indicator 3.7.2).

Another controversial issue is to provide vaccines and medicines without infringing on patent rights. Target 3.B elaborates on a position agreeable to all, to

> support the research and development of vaccines and medicines . . . access to affordable essential medicines and vaccines, in accordance with the Doha Declaration on the TRIPS Agreement and Public Health, which affirms the right of developing countries to use to the full the provisions in the Agreement on Trade-Related Aspects of Intellectual Property Rights (TRIPS) regarding flexibilities to protect public health, and, in particular, provide access to medicines for all.

In other words, the international regulation of patent rights allows some "copying" of medicines, etc., in poor countries (WTO 2001).

Targets 3.C and 3.D hint at the means necessary to achieve the goal in the form of increased financing, workforce development and risk reduction.

To summarize, the goal states a number of health-related problems but offers little in terms of how to solve these problems. Increased financing and workforce development, mentioned at the end, are important means to develop health-care services to achieve the desired ends.

Goal 4: quality education

The aim of goal 4 is to "ensure inclusive and equitable quality education and promote lifelong learning opportunities for all". It is important for children (and others) to go to school, and it is important to develop the workforce, but there are also short-term costs, such as a reduction of the workforce and an investment to be made.

The first target talks about completing "free, equitable and quality primary and secondary education", while the following two targets talk about access to what comes before (pre-primary education) and after (tertiary education). Target 4.5 adds a concern with the elimination of disparities based on gender and vulnerability.

Indicator 4.A.1 asks for data on a number of factors which are important for the quality of schools: the proportion of schools with access to: (a) electricity; (b) the Internet for pedagogical purposes; (c) computers for pedagogical purposes; (d) adapted infrastructure and materials for students with disabilities; (e) basic drinking water; (f) single-sex basic sanitation facilities; and (g) basic handwashing facilities.

Two targets ask for international cooperation through scholarships for tertiary education (target 4.B) and through teacher training (target 4.C).

Three targets talk about the desired outcomes, literacy and numeracy (target 4.6), substantial increase in the number of youth and adults who have relevant skills (target 4.4) and knowledge and skills needed to promote sustainable development (target 4.7). The latter includes a detailed specification, "including, among others, through education for sustainable development and sustainable lifestyles, human rights, gender equality, promotion of a culture of peace and non-violence, global citizenship and appreciation of cultural diversity and of culture's contribution to sustainable development". This can be read as a summary of what the SDGs are about.

It is further specified in a comparatively detailed manner in indicator 4.7.1, which asks for data on the "extent to which (i) global citizenship education and (ii) education for sustainable development, including gender equality and human rights, are mainstreamed at all levels in: (a) national education policies, (b) curricula, (c) teacher education and (d) student assessment".

To summarize, the goal elaborates on the ends to be achieved but gives little indication on the means to achieve them, except for a mention of teacher training.

Goal 5: gender equality

The fifth goal is to "achieve gender equality and empower all women and girls". Gender equality is an issue which affects all the goals and is repeated in some of them, but since it is a social issue, I will discuss it in that context.

The first target states the general goal to "end all forms of discrimination against all women and girls everywhere". The following targets elaborate on some of the problems to be dealt with, such as "violence against all women and girls in the public and private spheres, including trafficking and sexual and other types of exploitation" (target 5.2); "harmful practices, such as child, early and forced marriage and female genital mutilation" (target 5.3); unpaid work (target 5.4); lack of "full and effective participation and equal opportunities for leadership at all levels of decision-making in political, economic and public life" (target 5.5); and universal access to sexual and reproductive health (target 5.6).

The last target will be measured in two very different ways. Indicator 5.6.2 asks for data on the "number of countries with laws and regulations that guarantee women aged 15–49 years access to sexual and reproductive health care, information and education". More difficult is indicator 5.6.1, which asks for information on the "proportion of women aged 15–49 years who make their own informed decisions regarding sexual relations, contraceptive use and reproductive health care".

The final targets address specific issues. Target 5.A is about "reforms to give women equal rights to economic resources", which refers to the economic dimension of the SDGs. Target 5.B is about "the use of enabling technology, in particular information and communications technology". Target 5.C talks about "sound policies and enforceable legislation".

To summarize, the goal states a general principle, with emphasis on a selection of important problems. It suggests some means to achieve this, such as laws which guarantee certain rights. The mentioning of issues of gender equality under other goals makes it reasonable to think of gender equality also as a means to achieve other goals, such as economic development.

Goal 6: clean water and sanitation

The sixth goal is to "ensure availability and sustainable management of water and sanitation for all". This is a major problem for people who live in extreme poverty, which contributes to other problems, such as the spread of diseases, reduced agricultural output and a reduction of the time available for productive work (i.e. spending much time on bringing home water).

The first targets state the general goal to "achieve universal and equitable access to safe and affordable drinking water for all" (target 6.1) and to "achieve access to adequate and equitable sanitation and hygiene for all and end open defaecation, paying special attention to the needs of women and girls and those in vulnerable situations" (target 6.2).

Target 6.3 connects with the environmental dimension of the SDGs by pointing to improved "water quality by reducing pollution, eliminating dumping and minimizing release of hazardous chemicals and materials, halving the proportion of untreated wastewater and substantially increasing recycling and safe reuse globally". Target 6.4 adds "substantially increase water-use efficiency across all sectors and ensure sustainable withdrawals and supply of freshwater to address water scarcity and substantially reduce the number of people suffering from water scarcity".

More general issues are addressed in two targets: "implement integrated water resources management at all levels, including through transboundary cooperation as appropriate" (target 6.5) and "protect and restore water-related ecosystems, including mountains, forests, wetlands, rivers, aquifers and lakes" (target 6.6).

The final targets bring in "international cooperation and capacity building" (target 6.A) and "the participation of local communities" (target 6.B).

To summarize, this goal has strong links to the environmental goals. It states the desired ends and suggests some means to achieve them, primarily management systems and protection of ecosystems.

Goal 7: affordable and clean energy

The aim of goal 7 is to "ensure access to affordable, reliable, sustainable, and modern energy for all". Like the previous goal, this is part of the social dimension, as well as the environmental dimension.

Target 7.1 states the general principle to "ensure universal access to affordable, reliable and modern energy services"; to be measured by access to electricity (indicator 7.1.1); and to have a "primary reliance on clean fuels and technology" (indicator 7.1.2). Lack of electricity is a problem similar to the lack of water and sanitation (noted earlier) since it makes it difficult to achieve other aspects of a decent life, including the development of farming, mentioned earlier.

The issue of clean energy is further mentioned in the following targets, which go on to ask countries to "increase substantially the share of renewable energy in the global energy mix" (target 7.2) and "double the global rate of improvement in energy efficiency" (target 7.3). These targets hint at the problems with the use of oil and coal, which is a major problem, especially in relation to increasing living standards.

The most controversial issue during the negotiations was the treatment of fossil fuels, which are not explicitly mentioned under this goal on energy. There is a target under goal 12 on the rationalization of inefficient subsidies for fossil fuels (target 12.C), which was the extent of agreement on this subject (Kamau, Chasek & O'Connor 2018:176f).

The final targets for this goal mention international cooperation and investment (target 7.A), as well as expanded infrastructure and upgraded technology in the developing countries (target 7.B).

To summarize, the goal focuses on clean and renewable energy but gives only a few suggestions on what needs to be done, apart from some support for developing countries. It has been criticized for not saying very much on the problem of subsidies for fossil fuels.

Goal 11: sustainable cities and communities

The final goal in the social dimension is goal 11: to "make cities and human settlements inclusive, safe, resilient and sustainable". This goal deals with social issues but is often seen as one of the environmental goals since it refers to the inhabited area – a complement to nature.

The first target is in line with the social goals by asking for "access for all to adequate, safe and affordable housing and basic services and (to) upgrade slums" (target 11.1), while target 11.2 adds a concern with transportation systems. Target 11.B talks about "support least developed countries, including through financial and technical assistance, in building sustainable and resilient buildings utilizing local materials". Target 11.4 adds a concern for the world's cultural and natural heritage.

Target 11.3 asks for an inclusive and sustainable process of urbanization, as well as planning and management, including participation by civil society (indicator 11.3.2). Target 11.A asks for "support (for) positive economic, social and environmental links between urban, per-urban and rural areas by strengthening national and regional development planning", and target 11.B furthermore talks about "substantially increase the number of cities and human settlements adopting and implementing integrated policies and plans towards inclusion, resource efficiency, mitigation and adaptation to climate change (and) resilience to disasters".

The environmental aspects are shown in target 11.5, which relates to disasters, including water-related disasters. Target 11.6 relates to the "environmental impact of cities, including by paying special attention to air quality and municipal and other waste management". Partly on the same note, target 11.7 addresses "universal access to safe, inclusive and accessible, green and public spaces, in particular for women and children, older persons and persons with disabilities".

To summarize, this goal is a bridge between the social dimension and the environmental dimension. It states a few ends to be achieved and suggests some means to do this, mainly by support for developing countries and strengthened planning.

Summary

The seven social goals can be read as illustrations of the specific problems which affect people who live in poverty, but some of them are clearly relevant for rich countries. Hunger, health and basic education are examples of the first kinds, while clean water and energy are of the second kind. The goals overlap with the other dimensions, where some goals highlight economic aspects (hunger) and some highlight environmental aspects (water, energy). Gender equality is a very general goal.

These goals are more focussed on ends than on means, especially in comparison to the goals on the economic dimension. This means that they state a number of concerns but add only a little in terms of how to achieve them. Some examples of means are food production and government planning. This, in turn, makes it reasonable to think of the economic goals as means to achieve the social goals. A main problem is to find the financial means to pay for social services and basic investments. At the same time, the social goals include means to achieve some of the economic goals – for example, in the form of education and skills development ("human capital") needed for advanced production.

Economic growth is a way to create resources which can be used to pay for education and health care, etc. Several goals talk about resource mobilization, but the more specific issue of domestic taxes is almost not addressed at all. It is most explicitly stated in goal 17 (below).

The environmental dimension

There are four goals which relate to traditional environmental concerns: goals 12–15. There are links from some of the other goals to environmental issues, as indicated earlier. I will come back to how these goals relate to the economic and social goals, if the environmental goals are limits on the others and if the others contribute to or hinder the fulfilment of the environmental goals.

Goal 12: responsible production and consumption

The aim of goal 12 is to "ensure sustainable consumption and production patterns". It is a bridge between the economic and environmental issues. It formulates key aspects of sustainable development by placing limitations on consumption and production, which are the two fundamental dimensions of the economy.

Target 12.1 states the general goal of working with these issues and asks the developed countries to take the lead, "taking into account the development and capabilities of developing countries". Target 12.2 specifies what this should lead to – i.e. to "achieve the sustainable management and efficient use of natural resources".

The following targets add goals on waste management (target 12.3), waste generation (target 12.5) and a concern with "environmentally sound management of chemicals and all wastes" (target 12.4). Target 12.6 relates to businesses and target 12.7 to public procurement.

Target 12.8 talks about ensuring "that people everywhere have the relevant information and awareness for sustainable development and lifestyles in harmony with nature". Indicator 12.8.1 is almost identical to indicator 4.7.1, with the difference being that it asks for "education for sustainable development, including climate change education".

The final targets talk about support for "developing countries to strengthen their scientific and technological capacity to move towards more sustainable patterns of consumption and production" (target 12.A) and to "develop and implement tools to monitor sustainable development impacts for sustainable tourism that creates jobs and promotes local culture and products" (target 12.B).

Target 12.C is the rather weak statement on fossil fuels mentioned earlier:

> Rationalize inefficient fossil-fuel subsidies that encourage wasteful consumption by removing market distortions, in accordance with national circumstances, including by restructuring taxation and phasing out those harmful subsidies, where they exist, to reflect their environmental impacts, taking fully into account the specific needs and conditions of developing countries and minimizing the possible adverse impacts on their development in a manner that protects the poor and the affected communities.

To summarize, the goal states an important principle of the SDGs. It contains a number of ends, but few means to achieve them. Information seems to be the main instrument, in addition to strengthened technological capacity in developing countries. Subsidies for fossil fuels stand out as a weak area of the SDGs (as in goal 7).

Goal 13: climate action

The following three goals cover the three main dimensions of the environment, on land and on water, as well as the climate. Goal 13 focuses on climate change to "take urgent action to combat climate change and its impacts". It makes a reference to the work on climate change within the UNFCCC – for example, what became the so-called Paris Agreement in December 2015 (UN 2015b). It was argued that the SDGs had to have a goal on climate change, even if there was a risk that this would limit the subsequent negotiations in Paris (Kamau, Chasek & O'Connor 2018:192ff).

The targets talk about the goals to "strengthen resilience and adaptive capacity to climate-related hazards and natural disasters" (13.1); "integrate climate change measures into national policies, strategies and planning" (13.2); "improve education, awareness-raising and human and institutional capacity" (13.3); and "raising capacity for effective climate change-related planning and management in least developed countries" (target 13.B). The target also makes a reference to the commitment of developed countries to mobilize $100 billion annually by 2020 for the developing countries (target 13.A).

To summarize, the goal is very short and general. It mentions "policies, strategies and planning" as well as "education, awareness-raising and institutional capacity". The more specific commitments were made in the Paris Agreement 2015 under the UNFCCC.

Goal 14: life below water

Goals 14 and 15 have more specific goals than goal 13, even though they are also the object of international negotiations in other fora. The targets set goals for 2020, 2025 and 2030.

The aim of goal 14 is to "conserve and sustainably use the oceans, seas and marine resources for sustainable development". Its targets talk about marine pollution (target 14.1), managing and protecting marine and coastal ecosystems (target 14.2), ocean acidification (target 14.3), harvesting and overfishing (target 14.4), conserving at least 10 per cent of coastal and marine areas (target 14.5), prohibiting certain forms of fisheries subsidies (target 14.6), increasing the economic benefits to small island developing states and least developed countries (target 14.7).

Target 14.A demands an "increase (in) scientific knowledge, develop research capacity and transfer marine technology", while target 14.B is about "access for small-scale artisanal fishers to marine resources and markets", and target 14.C talks about "the conservation and sustainable use of oceans and their resources by implementing international law".

To summarize, the goal highlights problems at sea and points to a number of means to deal with them. Conservation and protection are mentioned, as well as prohibiting certain subsidies.

Goal 15: life on land

The aim of goal 15 covers the situation on land – i.e. to "protect, restore, and promote sustainable use of terrestrial ecosystems, sustainably manage forests, combat desertification and halt and reverse land degradation and halt biodiversity loss". Like goal 14, it contains a list of specific goals.

Nine targets cover a variety of issues related to forests, wetlands, mountains and drylands (target 15.1). Target 15.2 addresses deforestation and target 15.3 desertification, drought and floods. Target 15.4 mentions mountain ecosystems, including their biodiversity. Target 15.5 talks about the degradation of natural habitats and the loss of biodiversity. Target 15.7 addresses poaching and trafficking of protected species of flora and fauna, while target 15.8 addresses invasive alien species.

Target 15.6 demands that governments "promote fair and equitable sharing of the benefits arising from the utilization of genetic resources and promote appropriate access to such resources, as internationally agreed". This hints, for example, at the use of genetically modified plants and seeds.

A general principle is stated in target 15.9, which demands that governments integrate ecosystem and biodiversity values into national and local planning, development processes and poverty reduction strategies and accounts. This relates to the economic and social goals (noted earlier).

Target 15.A talks about the mobilization of financial resources for conservation and sustainable use, while target 15.B specifically talks about forest management and target 15.C about combatting poaching and trafficking of protected species.

To summarize, goal 15 has a relatively long list of ends but only a few specific means to achieve them. Planning and resource mobilization are the key items.

Summary

The four environmental goals make up an area of their own, much independent of the other two dimensions: the social and the economic. The environmental goals contain a list of ends to be achieved, which are common in the general debate and relate to some key behaviour (production and consumption), as well as to nature in a broad sense (climate, land and sea). The specific goals on nature are in line with other international agreements. When it comes to human behaviour, the goals are such that most people can agree with them.

With a risk of simplification, one could say that the ends have popular support, while some hotly debated topics are absent, such as specific suggestions on lifestyles (for example, on the consumption of meat). Another criticism is that the text is weak on policies which have an obviously negative impact on the environment, such as the subsidies for fossil fuels.

There are few drivers and barriers mentioned in the environmental goals, and there is almost nothing that can count as a discussion about the causes of the problems. One could, of course, argue that there is an implicit cause in the form of overuse of environmental resources, but that doesn't take the analysis very far. The relevant and much more contested question is, why is there overuse of the environment? Some environmentalists argue that the problem is greedy human nature, amplified by the way the economy works ("capitalism"; Newell 2012), while others argue that the key is a lack of appropriate regulation to control human behaviour, as illustrated in the fable of "the tragedy of the commons" (Andersen & Libecap 2014). These suggested causes point towards very different remedies, which indicates that the fundamental disagreement needs to be addressed in order to design reasonable strategies.

Political instruments, such as planning and information, are suggested in these goals, as well as resource mobilization. There is nothing on more controversial instruments like a legal prohibition of certain activities or the use of instruments to make the polluters pay for their effects on the environment.

By implication, the economic goals and the social goals can be thought of as (partial) means to achieve these ends, rather than (only) being problems and restrictions on the environmental dimension. The economic goals (noted earlier) make references to environmental goals as restrictions on economic growth. The environmental goals make reference to economic resources as means to solve some of the problems (for example, through technological development). In other words, economic activity is a problem as well as a solution to the environmental problems. This is a fundamental dilemma to be handled through careful development of national policies. There is little explicit guidance in the SDGs on how to deal with the dilemma.

Cross-cutting issues

The final two goals (goals 16 and 17) highlight cross-cutting issues which can be interpreted as prerequisites to reach the other goals. They address issues of implementation, especially the role of governments and the role of international support.

Goal 16: peace, justice and strong institutions

The aim of goal 16 is to "promote peaceful and inclusive societies for sustainable development, provide access to justice for all, and build effective, accountable, and inclusive institutions at all levels". This goes beyond the economic, social and environmental dimensions by introducing factors which have to do with governments and politics. The implication is that these factors are important means to achieve the SDGs, rather than ends in themselves, although one can argue that they are important in themselves. As mentioned before, there was a suggestion that these issues should be seen as a fourth dimension of sustainable development (Kamau, Chasek & O'Connor 2018:201ff).

While there was much debate concerning this goal, there are, nevertheless, nine specific targets about issues such as violence (targets 16.1 and 16.2); the rule of law and justice for all (target 16.3); organized crime (target 16.4); corruption (target 16.5); legal identity for all, including birth registration (target 16.9); and public access to information and the protection of fundamental freedoms (target 16.10). A tenth target calls for broader and stronger participation of developing countries in the institutions of global governance (target 16.8).

Target 16.A talks about building capacity, with an indicator on independent national human rights institutions. Target 16.B talks about non-discriminatory laws and policies for sustainable development, with an indicator on discrimination prohibited under international human rights law.

To summarize, goal 16 introduces a number of factors which are relevant to the SDGs as a foundation for the other goals. It is not clear from the text what the causal relations are but, by implication, the message is that governments are important in the process of achieving the economic, social and environmental goals. Peace, justice and institutions are fundamental items in a strategy to achieve the SDGs.

Some of the other goals noted earlier make reference to things governments do. Goal 16 can be read as support for the general idea that good governance/state capacity is important for development and for policies in general (e.g. Levy & Kpundeh 2004).

Goal 17: partnerships for the goals

Goal 17 is another add-on to the main goals, stating principles of global cooperation, i.e. to "strengthen the means of implementation and revitalize the Global Partnership for Sustainable Development". There are 19 targets which cover several themes of support: finance, technology, capacity building, trade and some general issues, such as global macroeconomic stability.

Five targets relate to financial issues. Target 17.1 is very interesting in relation to the social dimension and other goals where governments are asked to provide services. The target talks about "strengthen domestic resource mobilization, including through international support to developing countries, to improve domestic capacity for tax and other revenue collection". Indicator 17.1.2 talks about funding by domestic taxes. Target 17.2 talks about the commitments by developed countries to development assistance, and target 17.3 talks about additional financial resources from multiple sources, such as foreign direct investment (indicator 17.3.1) and remittances (indicator 17.3.1). Target 17.4 addresses long-term debt sustainability. Target 17.5 relates to investment promotion regimes.

Three targets relate to issues of technology. Target 17.6 talks about knowledge sharing, for example through a global technology facilitation mechanism. Target 17.7 specifically addresses environmentally sound technologies. Target 17.8 talks about a technology bank and capacity-building mechanisms for science, technology and innovation.

One target deals with capacity building to support national plans to implement the SDGs (target 17.9).

Three targets deal with trade. The first relates to the WTO by asking for a universal, rules-based, open, non-discriminatory and equitable multilateral trading system under the WTO (target 17.10). The second talks about doubling the least developed countries' share of global exports by 2020 (target

17.11). The third talks about duty-free and quota-free market access on a lasting basis for all least developed countries (target 17.12).

General systemic issues are dealt with in seven targets. Target 17.13 calls for global macroeconomic stability, including through policy coordination and policy coherence, while target 17.14 asks for policy coherence for sustainable development.

Target 17.15 calls for respect for each country's policy space and leadership to establish and implement policies for poverty eradication and sustainable development. Target 17.16 focuses on multi-stakeholder partnerships that mobilize and share knowledge, expertise, technology and financial resources, while target 17.17 focuses on public, public-private and civil society partnerships.

Target 17.18 calls for capacity-building support to increase the availability of data (for the indicators of the SDGs). Target 17.19 talks about measurements of progress on sustainable development that complement gross domestic product.

A big debate on the last goal was how to deal with the previously agreed upon principle of "common but differentiated responsibilities" (Kamau, Chasek & O'Connor 2018:205:f). This goes to the core of the SDGs which apply to all countries, while obviously the available resources differ greatly among them. It is even more controversial when some developing countries are becoming richer than some developed countries. The concepts of developing and developed are beginning to lose their meaning as a typology of countries.

To summarize, goal 17 elaborates on support for developing countries. It talks explicitly about taxes and support for government capacity building, as well as respect for government autonomy. Other instruments elaborated on are technology development and trade. These are clearly a means to achieve the ends elaborated on in other goals.

Summary

The final two goals broaden the SDGs by bringing in a number of factors which are important in order to achieve the other goals. In other words, they specify a set of drivers (and barriers) for sustainable development. Goal 16 relates to the role of governments in society while goal 17 mainly has to do with support from developed to developing countries.

The national governments are important for all design and implementation of policies, but the factors mentioned have a wider scope. Justice and properly working institutions are often seen as values in themselves. They are also part of a theory on what drives economic development. It seems that the SDGs contain the elements of at least a partly elaborated theory of how to achieve the SDGs in countries which are very little developed and with large domestic problems (so-called failed states).

The international support relates to important issues to help developing countries catch up with the developed countries, such as technology development and capacity building. Interestingly, this includes the development of taxation, which is an important aspect of government capacity. This, in turn, suggests how policies to meet the social goals and environmental goals can be financed, which is a topic missing in the main goals.

How can we understand the goals as a strategy?

After going through all 17 individual goals, we can see that there is much of an implied theory on how to achieve sustainable development. The goals form a mainstream theory or causal model, which most people can agree with, as one would expect from a global declaration formed through a consensus approach. The general message of the SDGs is that the three major dimensions (economic, social and environmental goals) are largely compatible, while acknowledging that the environmental goals put restrictions on economic growth.

The general strategy

In summary, the SDGs are to a large extent a pro-growth theory of how to deal with poverty and the environment. The SDGs promote economic change and give an image of poor countries following in the footsteps of developing countries, such as China, or developed countries, such as the European countries. The SDGs are not about limiting economic growth or conserving lifestyles and traditional values.

The implied theory is that economic growth is an important means to achieve the social goals (welfare) and even the environmental goals (Figure 4.1). Economic growth will produce economic resources, including new technologies, needed to promote the social and environmental goals. There are also a few examples of a reverse relationship, where the social and environmental goals contribute to the economic goals, such as the promotion of education, which is an important contributor to economic growth.

The SDGs are more elaborate than this short summary and include some more specific suggestions on how to bring about economic growth and how to make the link from economic growth to the other dimensions. More importantly, the government plays key roles in the creation of economic growth, as well as in the transfer of economic resources into social and environmental outcomes.

A summary of the four dimensions (economic, social, environmental and governance) of the SDGs is the causal statement of a strategy inherent in the SDGs shown in Figure 4.2.

Growth (within limits) → Welfare + Environment

Figure 4.1 A brief summary of the strategy

Rule of law → Growth + regulation + taxes → Welfare + Environment

Figure 4.2 A summary of the strategy

Some government policies, summarized in terms of "rule of law", provide a foundation for economic growth. Taxation and regulation provide the means for financing welfare and protecting the environment. In this interpretation, all the main goals (economic, social and environmental) depend on the governance dimension (goal 16).

The general strategy can be further elaborated on by dividing it into three parts, two of which correspond to the final goals in the figure (social and environmental goals), and one that corresponds to the economic goals, which are intermediary in relation to the two other goals. The purpose of this elaboration is to see more clearly what contributes to the social and environmental goals.

The three partial theories are very general and do not take national contexts into consideration. A final step would be to adjust the causal model to the situation in each individual country.

The environmental goals

The environmental goals, as elaborated on in the SDGs, are mainly to be produced by technology transfer (targets 12.A and 14.A) and regulation of economic activity (targets 7.2, 12.C and 13.B). The importance of economic development and social development is implied rather than stated explicitly.

Economic development is a driver for the development of new technologies and important for the creation of economic resources to be used for protecting the environment. Here the emphasis is on the positive effects of economic growth rather than the negative effects. It is understood that

Economic growth + Environmental regulation → New technology → Environmental goals

Figure 4.3 Drivers for the environmental goals

there are negative effects to begin with, which need remedies. Regulation is one kind of remedy for the negative effects of economic growth, while technological development is another type. Economic growth contributes resources for both remedies and plays a great role in technological development (Figure 4.3).

Social development will lessen environmental problems based on poverty, though this is not explicitly mentioned in the SDGs. Hence, the social and environmental goals are mainly portrayed as parallel rather than interrelated goals.

The social goals

The drivers for the social goals are more vaguely stated. Taxation is only mentioned in goal 17, while regulation is mentioned in targets 2.B, 2.C, 3.C and 5.A. New technologies are mentioned in targets 2.A and 5.B. Only a few of the more specific factors needed to reach the goals are mentioned, such as teacher training (education) and room to manoeuvre within the patents for medicine (health care).

It is obvious that economic growth is an implied driver for social development, as it will create the economic resources needed to pay for social investments. There is also an element of social regulation in the SDGs, such as goal 10 on equality and goal 5 on gender equality (Figure 4.4).

The environmental goals are mainly restrictions in this context. They don't contribute to the social goals but rather make it more difficult to reach them by limiting the pursuit of the economic goals. In a way, they can be described as safeguards so that the pursuit of economic and social goals will not be self-destructive with severe consequences on the environment. This is a classic case of conflicting goals, where the SDGs are about promoting both types of goals: promoting social development within environmental limits.

The economic goals

Economic growth is a means rather than an end within the SDGs. One could argue that material resources are important for an individually fulfilling life, something which is hinted at in the discussion about economic discrimination. The social and environmental goals are higher ends and partly restrictions on growth.

Economic growth + regulation + taxation → New technology → Social goals

Figure 4.4 Drivers for the social goals

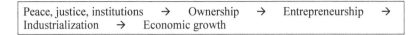

| Peace, justice, institutions → Ownership → Entrepreneurship → Industrialization → Economic growth |

Figure 4.5 Drivers for the economic goals

Some of the social goals can also be drivers for economic development, at least after some initial development. This goes for education and health care, which contribute to the development of the workforce. Social protection reduces some of the risks related to work, such as the risk of unemployment.

A number of factors – economic and political – are introduced as means to achieve growth. The economic factors include issues such as ownership (ending discrimination), entrepreneurship and industrialization. Trade is described as a driver for growth and is probably also important for raising the productivity of the economy. The same goes for technological development.

Regulation in the form of peace, justice and institutions are prerequisites for some of the economic drivers. Peace is of obvious importance, and the rule of law is often described as a key factor, meaning that governments are predictable and under control. Other targets, such as the reduction of corruption, have similar effects. The role of democracy ("inclusive institutions") is more uncertain; it's not a necessity for economic growth (below) but for a valued lifestyle with freedom of expression, etc. It may also be necessary for the legitimacy of regulation to achieve the social and environmental goals (Figure 4.5).

The governance goals

Governance is of great importance in the SDGs to achieve the other goals. There is a clear ambition to regulate the economy in order to achieve the environmental and social goals. Taxation is also important, especially for the social goals. Hence, peace, justice and institutions are important for all goals. Governance (the government) is important for activities mentioned in the SDGs, such as taxation, regulation, planning, conservation, awareness rising and other issues, including promoting trade. These activities can collectively be referred to as state capacity (Figure 4.6).

A major problem is that many governments have weak state capacity, being unable to carry out the activities which are needed to fulfil the SDGs. In a short perspective, humanitarian services can be provided by the international community in the form of aid. In a longer perspective, it is important that all governments develop state capacity, including accepted (legitimate) systems of taxation, to raise the funds for social and other services.

Peace → Rule of law → State capacity (= taxation, regulation etc.)

Figure 4.6 Elements of a theory of state capacity (good governance)

Democracy may increase state capacity under some circumstances, but it isn't a necessary prerequisite. State capacity can be upheld by authoritarian regimes, too, especially if they have a long-term view of economic development (see Chapter 6). Some authoritarian regimes have been successful in arranging the transition to a situation of economic growth (e.g. South Korea and China).

Another route to state capacity goes through the rule of law ("justice for all") and abolishment of corruption, which is a major barrier in many societies. It is also relevant in ethnically divided societies, where sub-groups like clans or tribes fight over dominance, which is one major reason why states fail (Hyden 2013). Here the rule of law means that a common centre of power is established to limit the power of the groups.

The rule of law refers to the relation between ruler, parliament and courts, where the idea is that rulers must obey the law and work through the law. Historically, this developed in Europe independently of democracy – for example, when parliaments put a limit on the kings' right to issue new taxes. This was long before democracy was introduced, which made democracy a consequence rather than a cause of the rule of law (Piattoni 2001). This is important to note when "modern" institutions are transferred to less developed countries with other governance arrangements (below).

The rule of law is often lumped together with state capacity to form the concept of good governance, but there is reason to see them as partly separated, to think of the rule of law as something independent and causally important for – as well as a limit on – state capacity (Gordon 1999). Likewise, "good governance" puts an emphasis on "good", while there is a need to open up for a discussion about the causal processes which lead from a situation where barriers dominate (dictatorship, corruption) to a situation where governments contribute to the SDGs (Møller 2017; below).

Peace is a fundamental prerequisite for the rule of law and state capacity, but the causality is likely more complex. The rule of law contributes to making peace lasting. These very fundamental processes lie far back in time for European countries and are still debated. The modern debate is rather what lessons can be learnt by developing countries: Are there shortcuts which can be taken to establish law, order and, hopefully, democracy? Can the SDGs help in making weak states develop in these respects? The SDGs hint at important background factors, but don't spell out a precise theory of state capacity. The next chapter will spell out some alternative interpretations of the causal mechanisms involved.

References

Andersen, Terry L. & Libecap, Gary D. 2014: *Environmental Markets: A Property Rights Approach*, Cambridge: Cambridge University Press.

de Soto, Hernando 2000: *The Mystery of Capital: Why Capitalism Triumphs in the West and Fails Everywhere Else*, London: Bantam Press.

Glenn, John 2007: *Globalization: North-South Perspectives*, Abingdon: Routledge.

Gordon, Scott 1999: *Controlling the State: Constitutionalism from Ancient Athens to Today*, Cambridge: Harvard University Press.

Hyden, Goran 2013: *African Politics in Comparative Perspective*, Second edition, Cambridge: Cambridge University Press.

International Labour Organization (ILO) 2009: *Recovering from the Crisis: A Global Jobs Pact Adopted by the International Labour Conference at Its Ninety-Eighth Session*, Geneva, 19 June.

Kamau, Macharia, Chasek, Pamela & O'Connor, David 2018: *Transforming Multilateral Diplomacy: The Inside Story of the Sustainable Development Goals*, London: Routledge.

Kymlicka, Will 2002: *Contemporary Political Philosophy: An Introduction*, Oxford: Oxford University Press.

Lerner, Josh 2009: *Boulevard of Broken Dreams. Why Public Efforts to Boost Entrepreneurship and Venture Capital Have Failed: And What to Do about It*, Princeton: Princeton University Press.

Levy, Brian & Kpundeh, Sahr (eds.) 2004: *Building State Capacity in Africa: New Approaches, Emerging Lessons*, World Bank Institute, Washington, DC: World Bank (The International Bank for Reconstruction and Development).

Møller, Jørgen 2017: *State Formation, Regime Change, and Economic Development*, Abingdon: Routledge.

Newell, Peter 2012: *Globalization and the Environment: Capitalism, Ecology and Power*, Cambridge: Polity Press.

Piattoni, Simona 2001: *Clientelism, Interests, and Democratic Representation: The European Experience in Historical and Comparative Perspective*, Cambridge: Cambridge University Press.

Piketty, Thomas 2014: *Capital in the 21st Century*, Cambridge: Belknap Press.

Rawls, John 1971: *A Theory of Justice*, Oxford: Oxford University Press.

Rudra, Nita 2008: *Globalizaton and the Race to the Bottom in Developing Countries: Who Really Gets Hurt?*, Cambridge: Cambridge University Press.

Sumner, Andy 2016: *Global Poverty: Deprivation, Distribution, and Development Since the Cold War*, Oxford: Oxford University Press.

United Nations 2015a: *Transforming Our World: The 2030 Agenda for Sustainable Development*, Resolution adopted by the General Assembly on 25 September 2015, A/Res/70/1.

United Nations 2015b: *Adoption of the Paris Agreement: Proposal by the President*, Framework Convention on Climate Change, Conference of the Parties, Twenty-First Session, FCCC/CP/2015/L.9/Rev.1.

World Trade Organization (WTO) 2001: *Declaration on the TRIPS Agreement and Public Health*, Adopted on 14 November 2001, Doha WTO Ministerial, WT/MIN(01)/DEC/2.

5 Alternative strategies for sustainable development

The previous chapter identified elements of a strategy inherent in the SDGs. This chapter will elaborate on the strategy by showing some alternative views on these matters. I will identify the three major perspectives on each issue in a pure and somewhat simplified form. This will contribute to a better understanding of what the SDGs actually say and to understanding the support (or background) for the SDGs by contrasting them with alternative perspectives.

The following is at the same time an elaboration on the causal statements inherent in the SDGs. It will help bring out important assumptions and possible omissions in the SDGs.

How can we think about drivers and barriers for sustainable development?

The SDGs cover many specific issues which can be summarized into four themes or major questions, as in the previous chapter. The questions will be presented below and then discussed throughout this chapter. For each of the questions, we can summarize the major debates in three political positions, which are similar across the issues. Models from political economy will help us get an overview of world views and assumptions in the political and scholarly debate. This will give us an overview of how we can think about drivers and barriers for sustainable development.

Four questions

There are four fundamental questions which need to be addressed by national governments, which relate to the four dimensions of the SDGs – i.e. the same as discussed in the previous chapter. I will begin with the question of economic growth (poverty reduction) and then discuss how to balance growth against the other dimensions of sustainability – i.e. social and environmental concerns. This order is reasonable when the social and environmental concerns are seen as higher ends for which economic growth is a means, or as balancing forces to economic growth. This is how the relationship is presented in the SDGs.

It is also convenient to begin with economic issues where there are a great number of theories, some of which encapsulate environmental concerns.

In the following, the first question is how economic growth can be pursued in poor countries (within social and environmental limits, to be further discussed in the second question). There are several perspectives and theories to guide such an inquiry, partly based on a discussion about what the poorest countries can learn from countries which have managed to get their economies to grow. The theories can be used to sort out the basic alternatives which are discussed globally.

The second question is how to balance the pursuit of economic growth against social goals while the third question is about growth and environmental goals. As indicated earlier, the concept of sustainability contains these three dimensions, which can be interpreted and pursued in different ways. Much of the international debate on the SDGs is about finding clever win-win solutions for these ambitions. The tensions and opportunities involved need to be spelled out in order to deal with the full set of SDGs. This is the case in rich as well as in poor countries.

The fourth question digs deeper, as mentioned in the previous chapter, by asking what the important background conditions for economic growth (and other goals) are. Economists, political scientists and others have recently focussed on the role of the state (governance) in laying the foundation for economic and social development. The concept of "good governance" has been a key recommendation by the World Bank and other donor organizations for several decades. Major problems stem from weak or even failed states, as well as from states which are captured by individuals or groups.

The four questions are obviously related. They are analytically distinct but most likely interrelated in the operation of national governments. Some aspects can be handled individually, while other aspects need to be dealt with in a coordinated manner. Governments need to think about the short term as well as the long term.

Three stylized perspectives

The SDGs can be compared to perspectives or theories of development to see what the inherent strategy in the SDGs is similar to and what it differs from. There are a multitude of perspectives in the academic and popular debates, which can be summarized in terms of three economic ideologies: liberalism, nationalism and Marxism. These are perspectives developed within International Political Economy (IPE), which is an academic discipline at the intersection of international relations, economics and political science (Jackson & Sørensen 2016).

IPE analyzes the politics of international economics by means of three main perspectives, which is what I will apply to the SDGs. The core idea is that there are liberal, nationalist and Marxist perspectives on the economy. Politicians and scholars see different problems and solutions, depending on which perspective ("paradigm") they belong to. These are, in other words, three conflicting political world views, with different policy suggestions, based on differing interpretations of the world and – to some extent – differing values (Jackson & Sørensen 2016; Miller 2018; O'Brien & Williams 2016; Figure 5.1).

The perspectives are so-called ideal types, meaning that they are analytical abstractions. They are pure and somewhat simplified to capture fundamental positions, while in reality there can be combinations of positions of various kinds. The three pure positions are a heuristic device intended to help us see the bigger picture without getting lost in too much detail. This, of course, entails a risk of overlooking important complexity and details, as well as empirical proofs.

I will use versions of the ideal types to describe positions in the debates on all four questions, which is also to say that they are alternative perspectives on the SDGs. The ideal types have different views on what the problems are and how they are caused, as well as differing views on what the solutions should be (and why). As a general introduction, they can be described as statements on the role of markets and states as coordinating devices in society. In other words, markets and states are shorthand descriptions of drivers and barriers to reach the SDGs.

Liberalism has a positive view of the market as a coordinating device, believing that markets allow individuals to act freely and to interrelate with each other as they please, and that this will lead to outcomes which are beneficial (efficient) for society at large (Jackson & Sørensen 2016:162ff;

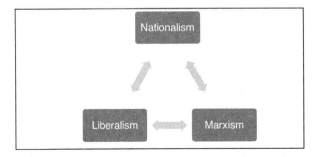

Figure 5.1 Three stylized perspectives

Miller 2018 chapter 2, O'Brien & Williams 2016:12ff). A characteristic description is that a market is like a computer which coordinates information based on the preferences of buyers and sellers, or, more to the point, that it is a decentralized system of coordination which relies on a minimum of centralized information. The market is seen as an ongoing process, striving to reach an equilibrium which it will never reach (but often described in textbooks as a state of equilibrium). The perspective is internationalist, believing in trade and other forms of open exchange across borders.

Nationalism, in this case, is a position which focuses on the interests of nations (states) and their competitiveness in relation other states. In the debate on the international economy, this perspective is often referred to as mercantilism. It sees a need for government intervention to improve the workings of the market (Jackson & Sørensen 2016:160ff; Miller 2018 chapter 4, O'Brien & Williams 2016:8ff). It identifies a number of failures where the market coordination is inefficient – for example, when costs (like pollution) are turned over to third parties, known as external effects. The nationalists are statists in the sense that they argue that governments (states) are better at solving these problems – for example, by introducing certain regulation.

Marxists are more critical of the market as a coordinating device, often based on the view that there is an unequal relationship between various parties in a market (Jackson & Sørensen 2016:165ff; Miller 2018 chapter 6, O'Brien & Williams 2016:16ff). The key term is exploitation, where they argue that rich people exploit poor people. Examples are that employers are usually stronger than employees in a bargaining situation since the employee is more dependent on a salary than the employer is on the individual employee. Collective bargaining is one way to deal with the situation. Government policy is another way. Similar arguments are made about the relationship between rich and poor countries; a radical transformation of the international system is necessary before poor countries can develop further.

These descriptions imply that there are interesting similarities and differences across the positions. Nationalists share some of the optimism over markets with liberals, and they also share some of the optimism over states with the Marxists. The liberals and Marxists share an internationalist leaning, which differentiates them from the nationalists.

The descriptions of the perspectives are mainly related to particular situations, which implies that it is an empirical matter to find out who is right about these issues. However, a problem is that there will seldom be agreement over the facts or even the definition of what constitutes a political problem. Another problem is that the concepts used to describe the situation are essentially contested (Connolly 1983). In this situation, the general models help us by providing a variety of interpretations of important facts and causal relationships. The perspectives define the situations differently and will come up with counterarguments to each other. The debate will never be settled. One advantage of discussing three ideal types is that it helps us understand ongoing debates.

Supporting arguments can be found in economics as well as in sociology, as two kinds of theories of human behaviour. Economics explains human behaviour as driven by incentives and consequences of actions, while sociology explains it as driven by norms and the legitimacy of action (March & Olsen 1989). Both types of theories are relevant in a discussion about what the problems and solutions are. Both types of theories can give support to the three perspectives of IPE.

There are some values involved in the IPE perspectives. The perspectives put different weights on individual and collective decision-making. There are also shared values with different interpretations – for example, when liberals and Marxists talk about making the individual free of exploitation. Their views of exploitation and freedom differ (below). I will come back to some examples of value-based arguments to support the IPE perspectives.

Political debates

We can think of the three positions as the core of three discourses on the SDGs (e.g. Clapp & Dauvergne 2011). In their pure forms, they emphasize different drivers and barriers, which is to make different assumptions about the causal factors to reach sustainable development. Liberals focus on the role of individual effort, private companies and technological development. Nationalists emphasize the role of the government in regulating and guiding behaviour. Marxists are more critical of the SDGs and see a need for remaking the international economy and controlling the greedy behaviour of individuals.

The three positions are close to political ideologies. Liberalism and Marxism are ideologies, while nationalism is close to social liberalism and socialism. What is absent is conservatism as one of the classical ideologies, referring to a perspective where the individual is of reduced importance. Traditionally, conservatism was about the importance of tradition and other collective entities (Heywood 2012). Some forms of environmentalism can be described this way, which I will come back to.

Another link to politics is with so-called epistemic communities. This is a perspective where politics is seen as driven by conflicts between ministries and their related groups (Haas 1992). For example, people involved in trade policy think in a different way than people involved in environmental policy. Trade policy tends to lean towards liberalism, while environmental policy tends to lean towards nationalism or even Marxism. While there are exceptions to this rough characterization, the observation is that policy communities are united by their thinking (episteme) – i.e. their interpretations of the world. Politics becomes a struggle over the interpretation of what is needed to achieve sustainable development – i.e. the causal assumptions.

Talking about a multitude of perspectives makes it easier to understand why actors talk past each other and why they try to build coalitions by focussing on common elements where the communities can find agreement across perspectives. The simplified positions open up a discussion about how politics develops and how the implementation of the SDGs is affected by a struggle for dominance in the national contexts. The framework can be used to understand the national implementation of the SDGs.

Drivers for economic development

Governments need to find a balanced way to achieve all the 17 SDGs, which entails finding a balanced way of pursuing economic, social and environmental goals. I will begin by looking at the debate on economic growth to begin a discussion about how it links to the other dimensions of sustainable development.

There are two questions I want to discuss. One is what are the drivers and barriers for economic development, or, more precisely, how we can think about these things? What perspectives are available in the discussion? The other question is what can the debate on economic growth contribute to the overall discussion about how the three dimensions of sustainable development should be handled?

The question of economic growth is close to what IPE deals with. It is fairly easy to elaborate on the model to relate it to this part of the SDGs (Jackson & Sørensen 2016; Miller 2018; O'Brien & Williams 2016).

The liberals would argue that economic development is the key to other kinds of development. They endorse many of the elements elaborated on in goals 8 and 9, such as entrepreneurship, investment and trade to drive industrialization. They are sceptical of the government as an instrument to guide the market beyond the establishment of rule of law. The liberal concern is rather with the many barriers where governments limit the forces of the market – for example, when the poor are discriminated against in matters of ownership (de Soto 2000; Moyo 2009). They emphasize the risks of corruption and oppression.

The nationalists also have a positive view on economic development, but they are more sceptical of the market than the liberals are. According to the nationalists, the government must take action to make the economy work properly (Woo-Cumings 1999; Chang 2003). One example in the SDGs is to provide infrastructure and electrification for the rural economy to take off. Another example is to invest in R&D to help firms and nations upgrade their technologies, which is a prominent theme in the SDGs. A third example from the SDGs is to invest in the workforce through education and health care. In either case, the state is seen as an instrument to correct weaknesses of market-based coordination (i.e. the liberal model), which means that the nationalists emphasize state capacity and see a need for upgrading in many developing countries, as suggested in goal 16.

The Marxist perspective is even more critical to the market based on general patterns of dominance in the national as well as the global context. Marxists see unequal relations where, for example, the poor and the environment are being used by multinational firms and other strong actors (Selwyn 2014; Wallerstein 2004). These inherent problems must be dealt with through drastic action, limiting human greed and exploitation. The Marxists want stronger action to be taken than just the SDGs. This means they have an even more positive view of the state as an instrument to pursue policies. The policies needed are almost revolutionary in asking for redistribution of power and economic resources, as well as changing human motives and personalities. There are few such radical elements in the SDGs, but there are targets on the power relations in international organizations, which would be endorsed by Marxists.

To summarize, the SDGs are closest to liberalism and nationalism on the issue of creating economic growth. The goals on the economic dimension contain elements that clearly belong to the mainstream debate on what is needed to develop economies, such as industrialization and trade. The goals also make a role for governments as regulators of the economy, which is more visible in the following dimensions.

Modernization theory vs dependency theory

The academic debate over economic development has focussed on a set of theories which has some relation to the three stylized perspectives. The debate over economic strategies in the 1950s and '60s was to a large extent a debate over modernization theory vs dependency theory (Williams 2012). The first argued that there was a common pattern of modernization which all countries needed to go through, which is largely up to each country to follow or not. The countries that failed were by implication seen as ruled by self-centred dictators more interested in enriching themselves than in pursuing long-term strategies for their countries' economic development.

Dependency theory shifted the focus to the role played by the individual countries in the global economy, where poor countries were seen as stuck in an unfavourable relationship with rich countries, along the lines of Marxism (noted earlier). It was argued that patterns of dominance and dependency were instituted during colonialism and that they remain in place today, though not as visibly as before. The rich countries and the multinational corporations have advantages which tilt the balance in their favour.

Modernization theory is compatible with textbook economics, where economic growth is the result of an ever-more efficient use of resources. Increased efficiency is driven by factors such as specialization, trade and economies of scale, as well as technological upgrading and skills development. It has elements of liberalism and nationalism. Innovativeness and regulation to support business activities are important, as well as investments in collective goods, such as infrastructure.

Modernization theory identified stages of development, which seem to be common for all countries. It is intuitively reasonable to think about developments in agriculture which create resources for industrialization, as was an important thought in the previous MDGs (UN Millennium Project 2005; Henley 2015). This goes together with an increase in education and training, as well as health and higher living standards. It is almost as if development is a self-driving process, if only certain barriers are removed and environmental aspects are taken into consideration.

Developmental states and global value chains

The two theories were followed by a new type of theory in the 1990s based on the experiences of the fast-growing countries in South East Asia. The model was referred to as a developmental state, focussing on the role of the state in the rapid modernization of countries like Japan and South Korea (Woo-Cumings 1999). A modern debate is whether China and India have followed the same model or not. Another question is if African countries can be understood through the same model (Whitfield et al. 2015). It has been argued that developed countries follow a similar model in relation to research funding, which is often a public support of technological development in large firms and elsewhere (Mazzucato 2013). The developmental state is clearly a nationalist theory in the model noted earlier, using the state to gain competitiveness through economic upgrading.

A key element is government capacity, since there is no guarantee that public subsidies for certain industries will actually lead to a strong international position. One hypothesis is that civil servants must be organized by a model of "embedded autonomy", at once understanding the needs of the firms (i.e. embedded) and able to avoid the short-sighted wishes of these firms (i.e. autonomous; Evans 1995). In either case, there is a strong belief in government capacity, which becomes an element of risk in countries where governments are weak.

Another factor in the explanation of rapid economic development is that these countries have produced goods for the global market. In other words, international trade and investments have been important ingredients. The key element is so-called global value chains, referring to the international supply chains where factories in poor countries can produce goods for the world market rather than just the national market. This is sometimes described as a shortcut to economic development (Glenn 2007).

A debate is on whether governments need to protect infant industries – i.e. new and vulnerable industries which begin the technological upgrading. There is also a debate on the value of foreign investments in poor countries and/or local production for the international market, whether these investments are mainly helping local economies grow faster or if they are examples of multinational corporations taking advantage of local resources, such as cheap labour and low environmental standards (O'Brien & Williams 2016:130).

The debate on what works in development is complicated by the heavy involvement of foreign aid. Much debate concerns the problems created by badly designed and executed policies (e.g. Easterly 2006; Hagmann & Reyntjens 2016). The dilemma is that aid can contribute much-needed resources, while it can also distort incentives, markets and government efficiency. The history of aid has gone through a number of waves which illustrate some of the problems (Williams 2012). The first wave, after the Second World War, was to invest in strategic collective goods, such as transport and electricity production. Disappointment with the lack of economic development led to a focus on improving government policy by means of structural adjustment programs. Eventually, these transformed into a focus on "good governance" as a prerequisite for efficient policies. The last wave of policy is the focus on MDGs and SDGs – i.e. on the ends rather than the means of the policies. In the next chapter, I will come back to one of the key debates.

Drivers for social development

The social dimension of the SDGs can be analyzed in a similar fashion, where the three perspectives can be used to show the main positions of the political and scholarly debates. Here I'm focussing on the general question of how to deal with poverty, which is close to – but not identical to – the previous question of stimulating economic growth.

Normative issues

The question of poverty was previously treated mainly as a normative issue of various conceptions of global justice, human rights, etc. Much of the debate was on how to treat poor people, especially if they have a right to a decent life and if rich people have a corresponding duty to help (e.g. Sen 1999). Aid was in this perspective thought of as morally obliged redistribution rather than a strategic intervention to kick-start the econo-mies of poor countries. The success of the countries in South East Asia led to a refocus on what works in development and what could be learnt from the Asian countries to lift other poor countries out of their situation (Woo-Cumings 1999; Henley 2015).

The successful development of many countries has led to a variety of national social policies in poor countries and an academic interest in analyz-ing how they developed, what their strengths and weaknesses are and how they are affected by globalization (Rudra 2008). It has also led to a debate on where the remaining pockets of poverty are located and what is needed to deal with them (Chandy, Kato & Kharas 2015).

Interestingly, there are no explicitly normative arguments in the SDGs. The goals are rather focussed on empirical issues, stating goals which con-tain elements of a strategy. The SDGs are an implied theory about empirical matters rather than normative issues. The SDGs are compatible with alter-native normative views, though they are more in line with the liberal notion of help to self-help rather than the Marxist notion of global redistribution. The normative debate comes closest in the issue of the role of the govern-ment in the economy and how to balance states and markets.

Three perspectives

The question of social development has several parts. The first is to design welfare systems, the services which should be covered and their level of ambition. General areas, such as education, health care and access to ser-vices like water and sewage, can be dealt with in many ways. The second part is to build capacity to deliver the services – for example, by training staff and making investments in needed facilities. The third part is how to fund these welfare systems by the individual client and/or by the govern-ment through taxation and other sources.

The liberal perspective would basically argue that economic growth by the market will trickle down to poor people, not as an automatic process but as business opportunities for an increasing number of people. Education (and possibly also health care) is seen as a tool for ambitious individuals to further their own interests and to get ahead. The demand for education will drive the supply of education, etc. Individual effort and innovativeness are the keys to abolishment of poverty, supported by an ethic of hard work (McCloskey 2010). Liberals have a negative view of governments as rais-ing barriers, discriminating against the poor or taking away the incentives for personal (and societal) improvement (de Soto 2000). Over time, liber-als have become supporters of limited welfare systems with incentives to encourage work and productive investments.

Nationalists see a greater need for government intervention – for example, to safeguard against the risks which are inherent in the economy. An argument is that individuals are ready to take on higher risks, including investing in education, if they know that there is support if they become unemployed or have an accident at work. It has been argued that this is in the interest of employers too and that it is difficult for them to organize such safety nets (Blyth 2002). All of this contributes to national competitiveness. Another argument in many developed countries is that universal welfare systems are the most efficient in terms of costs and support by the citizens (Goodin et al. 1999). A modern debate in some countries is that there must be a balance of rights and duties, meaning that you have to contribute payments to have a right to receive benefits from the social insurance systems.

Marxists have generally been in favour of government-funded welfare. Their opposition to early industrialization was much focussed on the immediate side effects on labour conditions and social affairs. The Marxist perspective on power asymmetries and exploitation lead to the implication that governments must take a larger role – for example, regulation to balance the profits and redistribution of wealth, nationally and globally. This is a position which has come to be accepted by most politicians in most countries, though there are also debates on the risks resulting from governments taking over responsibility for the consequences of individual behaviour (Machan 1987).

There is a line of argument which describes welfare as social investments, not as redistribution. The argument is that governments should invest in health care and education to gain a better workforce in the future (Heckman & Mosso 2014). This is interesting as a combination of several perspectives. Liberals can be attracted by the idea of making strategic investments, while nationalists like the increased competitiveness and Marxists like the publicly provided social services.

To summarize the social dimension, the SDGs are closest to nationalism and what was originally Marxism (i.e. public responsibility for welfare). The SDGs talk about the services that should be provided without saying much about the form or the financing. The differences between the three perspectives may surface at later stages of development. When it comes to the financing, taxation is mentioned towards the end of the SDGs. It is reasonable to conclude that the SDGs presume some form of domestic revenue to reach beyond immediate humanitarian needs. This, in turn, points towards the role of the government in raising revenues and providing services.

Drivers for environmental development/protection

The environmental dimension of the SDGs can be analyzed in a similar fashion, though the environmental dimension is linked to the economic dimension in two ways, with economic growth as a cause as well as a remedy for the environmental problems.

Two relationships

The first relationship between economic growth and the environmental situation is that the environmental problems to a large extent are side effects of industrialization and modernization. A very basic understanding of environmental problems is that they are caused by population growth and industrialization. Ever-more people who live at a high standard of living create demands on the environment (Speth & Haas 2006).

At the same time, advanced industrialization can produce new technologies with less impact and improved living conditions for the poor – for example, through the development of solar panels or better crops (Stern 2007, Chapter 16). Hence, economic development contains elements which could remedy some of the problems caused by other elements of economic development. Technological development was to previously build on fossil fuels, while it is now also about avoiding fossil fuels.

This points to a need to understand fundamental drivers, the underlying structural factors which are the causes of environmental problems and/or solutions. When will the economy deal with its own side effects, and when is government intervention necessary? Under what conditions is economic growth a driver for environmental protection, and when is it not? Some advanced economies are able to have growth as well as a reduced impact on the environment; are the adverse effects "outsourced" to other countries, or are they actually reduced?

Three perspectives

Liberals would argue that environmental problems are due to weaknesses in the system of ownership and other regulation. If ownership worked perfectly, pollutants would be charged and would need to take side effects into account. Liberals find the fable of "the tragedy of the commons" illustrative, where a common land is overused because it is collectively owned (Anderson & Libecap 2014). The market-based solution is to divide the commons into individual lots, where each owner decides how many cows he wants to feed. The problem is not the market but rather the lack of a market in the first place. Liberals also point to failed government policy (López & Toman 2006).

More advanced versions would argue that a market-based system drives innovativeness to come up with technical and other solutions to problems. The profit motive is important for individuals and firms in the development of clean technologies, etc. Liberals believe in green growth and ecological modernization – i.e. growth with a green profile, producing new technologies and more resource-efficient production (Hajer 1995).

Nationalists agree with the need for market-based solutions but take other lessons from "the tragedy of the commons". The need for new regulation is something that governments can take care of. Governments can also deal with situations where collective resources can't be divided up. The government can function as a club where collective decisions are made. Elinor Ostrom described how such collective solutions could form and work spontaneously under certain circumstances (Ostrom 1990). Here governments are needed to speed up and make the process more efficient. Governments need to introduce quasi-markets when real markets fail to emerge (Stern 2007). Nationalists could also say that common regulation is a way to guarantee fair competition in the marketplace.

Marxists are much more critical and see the market/capitalism as a root cause of the problems and something which can't be transformed by governments or new regulation. The exploitation of nature is similar to the exploitation of poor people and poor countries. The root cause of exploitation is human greed, which is fostered in a market-based system (Newell 2012).

Jennifer Clapp and Peter Dauvergne (2011) have developed a similar set of perspectives to understand the debate on environmental issues. In their version, only the liberals have the same name. The nationalists are called institutionalists to highlight their belief in regulation (institutions). On the more critical side are leftists (Marxists) as well as people who see the environment as a good thing in itself, regardless of its impact on the economy.

The fourth perspective has some similarities with premodern conservatism, which wanted to preserve traditional collectives and saw no value in economic development (which the other three perspectives do; Heywood 2012). In this version, conservatives end up on the same side as Marxists, united by their critical view of individuals and markets.

It is reasonable to conclude that the SDGs are mainly nationalist when it comes to the environmental dimension. The approach of the SDGs is to regulate markets to achieve better outcomes. The nationalists share the high hopes of markets with the liberals but see a stronger role for governments to deal with problems. This implies a great need for efficient governments to drive the policies.

Drivers for the development of governance/state capacity

The governance dimension of the SDGs (goal 16 and 17) can be analyzed similarly. Here the focus is on what kind of regulation (institutions) is necessary and can be provided in the various perspectives. It is about the framework conditions for economic growth (and other goals).

Liberals see a need for some framework regulation of the market. Their emphasis is on "the rule of law" as a minimum set of regulations to make markets workable and to keep governments limited – i.e. to avoid the risks with government interference in the market (de Soto 2000). Key rules for a market are about ownership, contracts, investment and trade. Courts and the rule of law are important to solve conflicts in society. Key rules for the government are to keep it limited and predictable, with constitutional limits and protection of civil liberties (Acemoglu & Robinson 2012). Democracy is seen as a system of limited government. Civil society must be strong to make the state civilized. Taxation is limited in the liberal perspective.

Nationalists see a bigger role for the state as an instrument to modernize society. The basic issue is to develop state capacity to play a larger role in society. While "the rule of law" is important as a procedural restriction on government, the most important aspect is the output: that governments control the economy, develop welfare systems and develop environmental regulation. An important aspect is to develop systems of taxation to raise the funds for services, including social investments (Moore, Prichard & Fjeldstad 2018). Nationalists aim for a high tax equilibrium, where citizens pay high taxes and believe that they get high-quality services in a fair and legitimate way. They would also say that a properly working government (without corruption) leads to increased levels of trust in society – i.e. it is the government which makes society civilized, rather than the other way around (Rothstein 2011).

Marxists see the state as captured by dominant interest groups. They see a need to turn the state into an instrument for more egalitarian relationships, nationally as well as globally (Cavanagh 2002). It is also an instrument to foster new behaviour which is more collaborative and peaceful. The role played by the government is even more demanding than in the nationalist vision. Some see a need for an educated elite to control the state – for example, for environmental reasons (Eckersley 2004). Others see a need for global democracy where citizens have equal status without attention to national boundaries (Bray & Slaughter 2015).

To summarize, the SDGs are again closest to the nationalist position, with a strong belief in the role of governments to drive necessary change. Liberals have a more restrictive view of the state, while the Marxists want to go much further with a strong state to drive through necessary changes in society.

Failed states

A special problem is the situation in countries where governments are unable to uphold fundamental functions, such as law and order. A number of states seem trapped in a situation where various groups compete over power and refuse to avoid violence (Lake 2016; Scott 2017). Many failed states are divided by social groups such as clans (Weiner 2013). Some kind of "law and order" can be upheld by warlords and other local leaders.

Another situation is where states exist but are captured by certain elites, ethnic groups or dictators (Koechlin & Förster 2015; Hyden 2013; Saylor 2014). This kind of rule is often some kind of "kleptocracy", where the ruling parties enrich themselves rather than upholding an impartial "rule of law". It can under some circumstances be combined with policies for development, which I will come back to in the next chapter.

Efficient states have taken control over groups and operate through the concept of rule of law (Fukuyama 2004). In other words, governments control society and are in turn controlled by society. This highlights the role of a rational legal system, but it doesn't explain where it came from or how to introduce this in countries where leaders are unwilling to modernize.

The pressing question in relation to the SDGs is what can be done to establish peace, rule of law and a government with state capacity. A standard explanation is that modern states grew out of older states, which were established by kings to fight wars and to raise taxes to pay for those wars (Tilly 1990). In a way, efficient states in Europe are the side effect of its long history of wars for dominance. It was partly driven by the high population density – i.e. the rivalry between ethnic groups. Africa was less densely populated, which made the forces for war and state-building weaker.

The implication for today's weak or failed states is that there is a need to move from a clan-based or ethnically based structure into a state-based structure operated through a legal system (rule of law; Weiner 2013). It is very difficult for clans to deal with the challenges implied in the SDGs, such as building infrastructure and developing a modern economy with social systems and environmental safeguards. The SDGs make no exceptions for premodern societies around the world.

It is also necessary to move from captured states to well-functioning states under the rule of law, which can develop state capacity further. The modernization of societies and economies is very difficult unless the state operates as an independent force for change.

It is very difficult to establish well-functioning states out of situations of civil war, ethnic rivalry or massive corruption. There is a long history of democracy promotion by organizations like the UNDP, the EU and individual states and non-governmental organizations. An optimistic reading of history is that environmentalism can contribute to the building of state capacity in underdeveloped countries (Death 2016).

In the next chapter, I will dig deeper into two strategies for developing state capacity.

Summary and implications for developing countries

The three perspectives have illustrated alternative ways to think about economic, social and environmental development, as well as the role of the government. Looked at it this way, the SDGs are generally taking a nationalist position, embracing much of what the liberals agree with. The SDGs allow a large role for the market and for governments to control and develop more efficient policies. The Marxist position is generally absent from the SDGs but important in the general debate on globalization.

The need for state capacity is clearly shown in the social and environmental dimensions. Governments are necessary to raise taxes to supply welfare services. They are also necessary to increase the concern with environmental side effects of economic growth. The lack of efficient governments in many poor countries is a major barrier for economic and social development, as well as for taking measures to protect the environment. Weak state capacity and/or corruption is a problem in more advanced economies too, especially where there are many social and environmental problems to deal with.

The SDGs can provide general support to deal with complex issues, including the development of better governments. The strong support for economic, social and environmental issues could widen to include support for rule of law and state capacity too. Environmentalism may give new energy to the general project of developing and modernizing societies around the world (Death 2016).

Even if environmentalism is about making sacrifices, putting restrictions on individuals and companies, it could be combined with liberal and nationalist values. Liberals believe in effort, responsibility, collaboration and equal treatment. Nationalists believe in collective action to promote common aims, including regulation to make markets more efficient to contribute to growth, welfare and environment. This could even give renewed support for what is called the Liberal World Order, an order built after the Second World War to promote democracy and market economy globally. The SDGs may be a new version of the Liberal World Order, with the potential of winning the hearts and minds of people in rich as well as poor countries.

References

Acemoglu, Daron & Robinson, James 2012: *Why Nations Fail: The Origins of Power, Prosperity and Poverty*, London: Profile Books.
Anderson, Terry L. & Libecap, Gary D. 2014: *Environmental Markets: A Property Rights Approach*, Cambridge: Cambridge University Press.

Blyth, Mark 2002: *Great Transformations: Economic Ideas and Institutional Change in the Twentieth Century*, Cambridge: Cambridge University Press.

Bray, Daniel & Slaughter, Steven 2015: *Global Democratic Theory: A Critical Introduction*, Cambridge: Polity Press.

Cavanagh, John, et al. 2002: *Alternatives to Economic Globalization: A Better World Is Possible*, A Report of the International Forum on Globalization, San Francisco: Berrett-Koehler Publishers.

Chandy, Laurence, Kato, Hiroshi & Kharas, Homi (eds.) 2015: *The Last Mile in Ending Extreme Poverty*, Washington, DC: Brookings Institution Press.

Chang, Ha-Joon 2003: *Globalisation, Economic Development and the Role of the State*, London: Zed Books.

Clapp, Jennifer & Dauvergne, Peter 2011: *Paths to a Green World: The Political Economy of the Global Environment*, Second edition, Cambridge, MA: The MIT Press.

Connolly, William E. 1983: *The Terms of Political Discourse*, Second edition, Oxford: Martin Robertson.

Death, Carl 2016: *The Green State in Africa*, New Haven: Yale University Press.

De Soto, Hernando 2000: *The Mystery of Capital: Why Capitalism Triumphs in the West and Fails Everywhere Else*, London: Bantam Press.

Easterly, William 2006: *The White Man's Burden: Why the West's Efforts to Aid the Rest Have Done So Much Ill and So Little Good*, Oxford: Oxford University Press.

Eckersley, Robyn 2004: *The Green State: Rethinking Democracy and Sovereignty*, Cambridge: The MIT Press.

Evans, Peter 1995: *Embedded Autonomy: States & Industrial Transformation*, Princeton: Princeton University Press.

Fukuyama, Francis 2004: *State Building: Governance and World Order in the Twenty-First Century*, London: Profile Books.

Glenn, John 2007: *Globalization: North-South Perspectives*, Abingdon: Routledge.

Goodin, Robert E., Headey, Bruce, Muffels, Ruud & Dirven, Henk-Jan 1999: *The Real World of Welfare Capitalism*, Cambridge: Cambridge University Press.

Haas, Peter M. 1992: "Introduction: Epistemic Communities and International Policy Coordination", *International Organization*, vol. 46, no. 1, pp. 1–35.

Hagmann, Tobias & Reyntjens, Filip 2016: *Aid and Authoritarianism in Africa: Development without Democracy*, London: Zed Books.

Hajer, Maarten A. 1995: *The Politics of Environmental Discourse: Ecological Modernization and the Policy Process*, Oxford: Clarendon Press.

Heckman, James J. & Mosso, Stefano 2014: *The Economics of Human Development and Social Mobility*, Working paper 19925, Cambridge: National Bureau of Economic Research.

Henley, David 2015: *Asia-Africa Development Divergence: A Question of Intent*, London: Zed Books.

Heywood, Andrew 2012: *Political Ideologies: An Introduction*, Fifth edition, Houndmills: Palgrave Macmillan.

Hyden, Goran 2013: *African Politics in Comparative Perspective*, Second edition, Cambridge: Cambridge University Press.

Jackson, Robert & Sørensen, Georg 2016: *Introduction to International Relations: Theories and Approaches*, Sixth edition, Oxford: Oxford University Press.

Koechlin, Lucy & Förster, Till (eds.) 2015: *The Politics of Governance: Actors and Articulations in Africa and Beyond*, Abingdon: Routledge.

Lake, David A. 2016: *The Statebuilder's Dilemma: On the Limits of Foreign Intervention*, Ithaca: Cornell University Press.

López, Ramón & Toman, Michael A. (eds.) 2006: *Economic Development and Environmental Sustainability: New Policy Options*, Oxford: Oxford University Press.

Machan, Tibor R. 1987: *The Main Debate: Communism versus Capitalism*, New York: Random House.

March, James G. & Olsen, Johan P. 1989: *Rediscovering Institutions: The Organizational Basis of Politics*, New York: Free Press.

Mazzucato, Mariana 2013: *The Entrepreneurial State: Debunking Public vs. Private Sector Myths*, London: Anthem Press.

McCloskey, Deidre N. 2010: *Bourgeois Dignity: Why Economics Can't Explain the Modern World*, Chicago: University of Chicago Press.

Miller, Raymond C. 2018: *International Political Economy: Contrasting World Views*, Second edition, Abingdon: Routledge.

Moore, Mick, Prichard, Wilson & Fjeldstad, Odd-Helge 2018: *Taxing Africa: Coercion, Reform and Development*, London: Zed Books.

Moyo, Dambisa 2009: *Dead Aid: Why Aid Makes Things Worse and How There Is Another Way for Africa*, London: Allen Lane.

Newell, Peter 2012: *Globalization and the Environment: Capitalism, Ecology and Power*, Cambridge: Polity Press.

O'Brien, Robert & Williams, Marc 2016: *Global Political Economy: Evolution & Dynamics*, Fifth edition, London: Palgrave Macmillan.

Ostrom, Elinor 1990: *Governing the Commons: The Evolution of Institutions for Collective Action*, Cambridge: Cambridge University Press.

Rothstein, Bo 2011: *Quality of Government: Corruption, Social Trust, and Inequality in International Perspective*, Chicago: University of Chicago Press.

Rudra, Nita 2008: *Globalization and the Race to the Bottom in Developing Countries*, Cambridge: Cambridge University Press.

Saylor, Ryan 2014: *State Building in Boom Times: Commodities and Coalitions in Latin America and Africa*, Oxford: Oxford University Press.

Scott, Catherine 2017: *State Failure in Sub-Saharan Africa: The Crisis of Post-Colonial Order*, London: I.B. Tauris.

Selwyn, Benjamin 2014: *The Global Development Crisis*, Cambridge: Polity Press.

Sen, Amartya 1999: *Development as Freedom*, New York: Anchor Books.

Speth, James Gustave & Haas, Peter M. 2006: *Global Environmental Governance*, Washington, DC: Island Press.

Stern, Nicholas 2007: *The Economics of Climate Change: The Stern Review*, Cambridge: Cambridge University Press.

Tilly, Charles 1990: *Coercion, Capital, and European States, AD 990–1992*, Oxford: Blackwell.

United Nations Millennium Project 2005: *Investing in Development: A Practical Plan to Achieve the Millennium Development Goals*, New York: Earthscan.

Wallerstein, Immanuel 2004: *World-Systems Analysis*, Durham: Duke University Press.

Weiner, Mark 2013: *The Rule of the Clan*, New York: Farrar, Strauss & Giroux.

Whitfield, Lindsay, Therkildsen, Ole, Buur, Lars & Kjær, Anne Mette 2015: *The Politics of African Industrial Policy: A Comparative Perspective*, Cambridge: Cambridge University Press.

Williams, David 2012: *International Development and Global Politics: History, Theory and Practice*, Abingdon: Routledge.

Woo-Cumings, Meredith (ed.) 1999: *The Developmental State*, Ithaca: Cornell University.

6 The governance challenge

A key to the implementation of the SDGs is the role of the government as an organization to plan and execute actions to bring about the SDGs in each country. This includes activities like regulation and taxation, as well as establishing the general rule of law and "inclusive institutions" – i.e. getting rid of corruption and biased governments. This is a general theme for all the goals, though it is specifically mentioned in goal 16.

These challenges are particularly strong for developing countries, which often face several of these challenges at once. Where should they begin? What comes first? Is good governance a prerequisite for development, as indicated in the SDGs, or is it actually a consequence of development? How important is state capacity and good governance to start development and modernization?

Good governance as the key to successful implementation of the SDGs

"Good governance" has been a key policy recommendation for developing countries for a long time. Organizations like the World Bank and national agencies for aid have insisted on seeing bad governance as a major problem for developing countries and a prioritized item to be dealt with. The recommendation has been part of the infamous Washington Consensus on aid policy, although sometimes seen as an addition and partly an alternative to it (Williams 2012; Serra & Stiglitz 2008).

The focus on governance in the aid policy community was at its strongest in the '90s and has been somewhat relaxed and to some extent replaced by a focus on general goals, such as the MDGs and the SDGs. As shown earlier, good governance is an important part of the SDGs, even though it's not among the most visible issues.

Good governance

The basic idea of good governance is that a properly functioning public sector (government) is more or less a necessity to deliver basic services in developing countries and is a framework for a market economy. This is the thinking that is found in the SDGs. Good governance is furthermore thought to strengthen trust in societies in general and to lay the foundation for more ambitious policies, where trust is essential, for example, to make citizens willing to pay taxes and make other sacrifices for the common good (Rothstein 2011). This is also relevant for the implementation of the SDGs.

The modern discussion uses the term "governance" to focus on the governing, the functions of public administration, rather than the government itself as one of several entities to perform those functions. Part of the solution in many developing countries is in the form of international support of various kinds. Under extreme circumstances, international organizations can provide services which governments are unable to provide. Over the long run, this capacity must be developed by each national government, as hinted at in goals 16 and 17, which mention peace, justice and institutions, including taxation.

The causal argument is that these arrangements lead to economic development and other goals, such as good health, education and environmental protection. Hence, the absence of good governance is seen by its proponents as a fundamental barrier for development, reducing the effects of all other efforts to stimulate development and a good life.

In addition to the causal argument, there is a value-based argument about good governance as a shorthand description of the only accepted design of government, at least in the global north. Good governance includes democracy, the rule of law, respect for human rights and so on. The definitions are usually very long and specific. My point is that they are defended variously on empirical grounds or normative ground, and that both types of argument are relevant in a discussion about good governance and the SDGs.

Disagreements

The issue of good governance has great relevance for global policies for development, such as the SDGs. Some proponents of good governance are critical of the global goals for not putting enough emphasis on good governance. An example is that issues relating to governance are treated as aspects of the last goals rather than highlighted as one of the key goals necessary to achieve other goals. By not emphasizing good governance, the other goals lose their backing by a causal argument. It is now up to each country to think about how they can achieve the goals and where they should begin taking action.

The debate on the SDGs is even more interesting since there are scholars who are questioning the causal role of good governance – i.e. arguing against what I call the inherent strategy of the SDGs (noted earlier). The argument is that there are other paths of development where at least some of the recommendations on global governance aren't necessary or even mistaken. There are other ways of development, exemplified by the rise of China, which has developed economically in spite of being an authoritarian state. This indicates that not all recommendations of good governance are necessary for economic growth. On the other hand, not all authoritarian states are developing, so there must be a more elaborate argument about what drives development and what blocks it.

In this chapter, I will analyze a specific proposal by Tim Kelsall and David Booth, together with Diana Cammack and Frederick Golooba-Mutebi (2010), that "developmental patrimonialism" is an appropriate strategy for developing countries. "Patrimonialism" is a term coined by Max Weber to describe premodern societies where the relationship between rulers and the ruled is based on personal loyalty, an exchange of protection for support. This is in contrast to modern forms of organization (public and private), where authority is based on formal roles – i.e. the rule of law (meaning the law is above the ruler) – and a bureaucracy is working to implement rules in a disinterested (impersonalized) way and leaders are held accountable by voters and other stakeholders.

Kelsall and Booth see patrimonialism as appropriate for development purposes. What they mean, to put it bluntly, is that some forms of corruption are beneficial for economic growth. By implication, corruption is also good for social development and the protection of the environment if we follow the inherent strategy of the SDGs (noted earlier). This implies that good governance may not be as important as others (including the founders of the SDGs) think and may even be a waste of resources, an inefficient strategy of development. This is a very interesting counter-point in relation to much of the standard arguments about good governance. The critical perspective could help identify the more specific circumstances under which development is likely to occur.

My ambition is to analyze the debate on good governance by focussing on the proposal by Kelsall and Booth. I will place their argument in context and begin a test of it. I will interpret their argument by spelling it out and showing what it implies.

Critical perspectives

Tim Kelsall and David Booth are interesting because they present a new theory of development with the ambition of reversing the dominant thinking about strategies for development. They build on other critical perspectives on good governance – for example by Ha-Joon Chang (2003a) – but they take a step further towards an alternative strategy.

The general criticism of good governance is summarized by Merilee Grindle (2004, 2007), who argues that the good governance agenda is too demanding. Instead, we should settle for "good enough governance", taking one step at a time, so to say. Rather than treating all countries the same way, we should think more specifically about appropriate steps for each country. If not, the more advanced goals will be difficult to achieve, and the actions taken may even make the situation worse. An example is where competitive elections lead to an increased effort by incumbents to buy votes, or where term limits make incumbents less interested in long-term solutions for the country. Appropriate advice on governance based on the experiences from developed countries may not be appropriate for poor countries.

Grindle argues, for example, that so-called failed states, without a functioning government, should focus on very basic problems and not try to solve more advanced problems before they have the basic functions of government in place. Similarly, countries with highly personal rule ("patrimonialism"), or only minimal institutionalization (only some aspects of democracy), should likewise be careful about the steps they take (Grindle 2007:564, based on Moore 2001).

In this version, the argument is basically that advice on good governance should take the context into account. This position is compatible with the SDGs and can be interpreted as a reaction to the simplest type of policy advice which recommends one type of action everywhere. Ambitious donor organizations are more nuanced and try to adapt their strategies to individual countries (Grindle 2007). However, this becomes more complicated when there is an abundance of organizations working to support individual countries, sometimes with clashing or even competing ambitions. The donor community as a whole is highly uncoordinated, for better or worse.

Implicitly, the strategies should build on empirical knowledge about what works. Implied also is that there are stages or even a pathway of development which provides guidance for policy. It would be prohibitively difficult to adapt recommendations to a context without a theory to base it on. Grindle's implicit theory is about stages of institutional development, while Kelsall and Booth question the pathway itself.

Developmental neopatrimonialism

Kelsall and Booth have elaborated on the argument of a more specific theory about economic progress in developing countries, especially in Africa. Their argument is that under some circumstances, corrupt (patrimonial) leaders will bring about economic development. I will come to the specific circumstances below. Their argument against a focus on good governance and in favour of a focus on economic development is that it is unrealistic (too demanding) and unnecessary to focus on governance (Kelsall & Booth 2010:26).

Like Grindle, Kelsall and Booth see some of the focus on governance as a wasted and even as a negative effort to improve the situation in developing countries. But in contrast to Grindle, they see positive effects of corruption and suggest that it should be accepted. According to them, certain forms of corruption can lead to some or possibly all the goals which Grindle and the proponents of good governance, such as the SDGs, want to achieve, but through another pathway which, the argument goes, is less demanding.

Kelsall and Booth see three advantages with developmental patrimonialism over good governance. One is the possibility of "primitive accumulation", creating a first set of resources to start off economic development. Another is the possibility of political stability for state-building and nation-building. A third is that it may put some discipline on corruption, which weak legal systems are unable to do (Kelsall & Booth 2010:7). All of these concerns are relevant enough to take into account.

Booth and Cammack (2013) provide an additional argument, which can be interpreted as a defence for pragmatism and room for leaders to go against popular opinion and even against demands for democracy. They see a focus on "principals and agents" as a restriction on aid policy. Principals and agents are part of an economic conceptualization of the relationship between rulers and the ruled. My interpretation is that Booth and Cammack find it wasteful that leaders should be held accountable by the voters. This goes against a major concern in the development community that leaders must be held accountable as a way to make them more interested in the needs of the people. In short, this aspect of democracy must stand back for the pursuit of economic growth and better living conditions. Checks and balances are seen as limitations rather than safeguards for economic policy.

Kelsall and Booth are not saying that developmental patrimonialism is necessary for development (though they seem to do that on page 13), only that it is an easier path than beginning with good governance. They also don't say that it is sufficient. They point out that developmental patrimonialism (the good kind of corruption) is not sufficient as a strategy for development. It has to be combined with broadly pro-capitalist policies (Kelsall & Booth 2010:26).

However, this is a difficult step in the argument. Tim Kelsall (2011) actually makes a more specific point, which can be interpreted several ways. He seems to agree with Grindle (2004) and Chang (2003b), who show that democracy and the general rule of law came after economic development in the Western world, while some specific aspects of governance came before, laying down some framework conditions for running a business and conducting trade (a market economy). Hence, the general argument seems to be that democracy and political reform can wait, while the basic conditions for a market economy cannot. It is not a total rejection of good governance, at least not for failed states or states with premodern legal systems – for example, legal systems based on religion rather than the principles of property rights.

Questions

The theory promoted by Kelsall and Booth actually has two parts, linking developmental patrimonialism to economic growth and linking economic growth to the development of good governance. The argument is not just that some forms of corruption are good for economic development, but that a focus on economic development will later lead to good governance. Hence, it is a better strategy in the sense that it deals with the standard of living first and secondly leads to an acceptable form of government.

The second part means that good governance is seen as a consequence of economic development rather than a prerequisite for it, as in the dominant perspective and in the SDGs. This is based on historical evidence that economic development started in the Western democracies before they had all elements of good governance. This is not only a problem for the theory of good governance but also a problem for good governance as policy advice. Ha-Joon Chang (2003b) talks about the rich countries kicking away the ladder for developing countries – i.e. denying them the strategy the developed countries pursued, making it more difficult for poor countries to catch up. The message is that the rich countries should let the poor countries develop the same way Europe did, not the way the ahistorical theories about good governance say they should.

It is not clear if Kelsall and Booth actually agree with the second part, though it seems important as support for their general argument. Kelsall and Booth are less certain about Africa. They say the best evidence is in Europe, while their examples of developmental patrimonialism have not yet developed into democracies.

This raises two major questions. Firstly, what is the evidence for the theory – i.e. that some forms of corruption can be a fast track for growth and that this will in turn lead to good governance (democracy)? Secondly, is it reasonable to put fundamental values like democracy on hold while we promote economic growth? In other words, what is the relationship between the empirical and the normative aspects of good governance? Is it reasonable to give up ambitions about politics (democracy) to prioritize growth? Should good governance in the form of democracy, rule of law and respect for human rights be promoted even if it means that growth is delayed?

Kelsall argues that better living conditions for the poor is a kind of human right too (Kelsall 2013).

The evidence

Kelsall and Booth (2010:12) find support in an overview of experiences from Africa. They use examples from various countries and at various times to show that developmental patrimonialism has two characteristics which make it successful: centralization and a long-term perspective. Countries develop economically when both are present.

The evidence presented by Kelsall and Booth is a typology where the machinery of government in a developing country can be centralized or decentralized, while the time horizon of the leaders can be long or short. Three of the combinations are not working well, while the fourth ("developmental patrimonialism" or "developmental rent utilization") more often leads to economic growth in the cases presented (Figure 6.1).

The combination of centralization with a short horizon is described as a "non-developmental kleptocracy", while the opposite combination is a "losing battle against corruption". The worst kind is an anarchy or a "free for all".

The strongest performance and strongest evidence for the theory are the cases of Cote d'Ivoire 1960–75, Malawi 1961–78, Kenya 1965–75 and Rwanda from 2000 (ibid.). There was almost as strong a performance in Ghana 1981–92, Uganda 1986–2000 and Malawi after 2004. All of these were ruled by strong leaders who centralized control (a monopoly on corruption) and used this control of the economy to diversify it (a long-term perspective). The leaders avoided the temptation to settle for short-term personal gain and opted for larger gains in the future, either for personal benefit or for the benefit of their countries. In short, the leaders were benevolent dictators or benevolent Mafioso.

Other data presented by Kelsall and Booth are more difficult to fit with the model – i.e. the successful cases are in the lower right-hand quadrant and the failed cases in the other three quadrants of Figure 6.1. Several countries

	Low centralization	**High centralization**
Short horizon	Free for all	Non-developmental kleptocracy
Long horizon	Losing battle against corruption	Developmental rent utilization

Figure 6.1 The typological theory

Source: Kelsall & Booth 2010:8

changed characteristics over time and/or their economic performance – for example, Kenya, where the time horizon has been short since 1980 with expectedly poor economic performance, but also with reasonable performance (after 2002). Ghana was interestingly a developmental patrimonial state twice, but with different economic performances, poor in 1957–66 and quite strong in 1981–92. Tanzania is another example of a developmental patrimonial state with a poor performance, which is explained by the lack of pro-capitalist policies – i.e. doing things which go against the development of the economy.

Kelsall and Booth use additional examples from Asia to show that their theories hold. The key example is Indonesia 1966–97 under Suharto, when it was one of the fastest-growing economies in the world. "The Indonesian economic model was wasteful, politically repressive, and environmentally destructive, but its growth potential cannot be denied", they argue (2010:5). Under Sukarno, the successor, Indonesia lost its long-term focus and degenerated.

China may be an even better example of the model at work, where the Communist Party to a large extent acts within the concept of developmental patrimonialism. South Korea is even more interesting since it pursued authoritarian policies for about 25 years and then developed into a democracy in 1987. Japan, in contrast, pursued developmental policies and good governance after the Second World War.

The typology can be read as a distinction among types of corruption and as a warning that the optimal combination can degrade into one or the other. The differences between them are matters of degree rather than kind. In other words, it is difficult to establish where the boundaries between categories should be drawn. This is important to address in a wider comparison of theories. Some of the successful examples given have in fact broken down over time.

More problematic is that not all the examples of developmental patrimonialism show performance superior to the other combinations of policy. Under some circumstances, the bad combinations have led to "reasonable" or "mostly reasonable" performance, while the good combination has led to "poor" performance as well as "strong" and "quite strong" performances (Kelsall & Booth 2010:12). The fact that some cases of developmental patrimonialism perform worse than the other combinations is a problem for the theory and indicates that there are more factors to take into consideration.

A further limitation on the analysis is that Kelsall and Booth don't look at all possible data. They give examples, most of which fit the theory, but leave it up to the reader to think about cases which may not fit. A more ambitious test would be to go through all countries to see if failure or success can be explained by the theory.

They also don't look at cases where good governance has led to growth or where it has stifled growth. There is no explicit comparison of the two theories: the traditional and their own. The argument by Kelsall and Booth is rather a falsification of a strong version of the theory of good governance, that it is the only way to bring about economic development. The successful cases of developmental patrimonialism show that it isn't, but we don't know the relative merit of the two strategies.

Differing methods

The typology is not a theory in itself but rather a means of classification. It is meant to show that certain characteristics of the state lead to certain economic performance. It is based on detailed description and comparison of the cases, in contrast to statistical testing of data. This methodological choice partly explains differences in the findings on governance arrangements from the theory of good governance.

Marilee Grindle argues that methodology has a great impact on research and on policy recommendations (Grindle 2004, 2007). The more general theories of good governance are based on quantitative studies which show general causal relationships, often through the work of institutional economists or some political scientists. The more specific theories making statements about individual cases are based on qualitative studies, illustrated by the typology noted earlier.

Proponents of good governance seem to rely on economics, pointing out causal links based on cross-country comparisons to find general relationships on a high level of abstraction. The critics apply qualitative methods, as found in sociology, which identify particular processes over time in more limited settings. Political scientists ("comparativists") work within both traditions.

Qualitative research makes important contributions, especially in the form of process tracing, describing how events unfold (Landman 2013). The statistical type of knowledge is of value to make hypotheses and for some kinds of testing, but it has to be compatible with the case studies. The cases provide the strongest evidence but with other risks of misinterpretation due to the low level of generalizability. From a methodological point of view, the effort by Kelsall and Booth makes sense – i.e. to look for cases which falsify the general theory. The method is reasonable, but the weak findings (low correlation between governance and economic performance) is a problem.

Two rival theories

One way to understand the debate better is to place it in the context of a major struggle between fundamental strategies for development or even between political ideologies. Kelsall and Booth make references to developmental states, the model discussed earlier, where states guide the economy to promote upgrading and structural change (Woo-Cumings 1999). As described earlier, the term was coined to describe countries like Japan, South Korea and China, where the state worked in partnership with businesses to promote technological advancement. This is often held as the explanation for the rapid development of agriculture, manufacturing and high-tech businesses. Proponents argue that elements of this model are applied in countries like Germany and the United States too.

The developmental state is mainly an instrument for economic upgrading in a top-down manner – i.e. a nationalist strategy in the vocabulary noted earlier. However, the government shouldn't make its decisions in splendid isolation, but must rather be closely connected with firms and work with them to find out what it should do. Its strength comes from picking winners and supporting them to be successful in the global marketplace. This is in contrast to the liberal strategy mentioned earlier, where states stay away from firms and rather uphold competition to make sure that the market selects winners from the bottom up.

The policy of good governance is closer to the alternative model, which emphasizes the role of markets and the need to put in place institutions which will guide actors to produce socially beneficial outcomes from their individual actions – for example, through competition. This is a more classical liberal view of markets, which is sceptical of the capacity of states to make things right. In reality, the difference may be a matter of degree since both want markets to function properly and both see at least some roles for the state. However, in terms of economic ideologies, one is liberal and the other is nationalist (see the aforementioned).

In the broader debate, there is some agreement that the two perspectives point at each other's weak spots. The most famous Asian developmental states managed to speed up the upgrading of the economy through planning and investments, but an experimentally oriented economy like the United States has advantages in finding new paths and taking advantage of decentralized information (e.g. Hall & Soskice 2001).

However, a major change since the early developmental states concerns the role of the national economy. Fifty years ago, several countries closed their borders to protect infant industries. Import substitution was the concept of the day. Now, the focus is rather on global value chains and the benefits of producing goods and services for a global market (Glenn 2007). This is seen as a shortcut to economic development, since the global market is much bigger than the national market. Hence, an argument is that developing countries should aim for economic integration rather than isolation. Integration into world markets will affect the policy choices.

Developmental patrimonialism may have some strengths as an export-oriented strategy, but it also has some limitations. Dictatorships may have some advantages for foreign investors, such as granting monopolies, but there will also be risks that dictators change their minds or lose their hold on power. Furthermore, it will likely put a limit on the transfer of knowledge from the investor to the local firms, not least in the form of spin-offs or service providers. These mechanisms are likely to be stronger if it is easy to start new firms. The strategy of good governance seems more compatible with the needs of foreign investors, such as a strong general business climate and a political system which is trustworthy and operates with checks and balances.

Are the strategies feasible?

Another way to understand the two strategies better is to look at their feasibility and, more specifically, what it takes to make them successful.

Beginning with the strategy of developmental patrimonialism, Kelsall and Booth already discussed possible risks in the form of degeneration of the model – i.e. the three alternative combinations in the matrix noted earlier. The optimal model can become short-sighted and/or the centre may lose its grip. Several of their national examples have degenerated. At the same time, they see the possibility of a positive development, too: the model evolves into an improved governance, even what the proponents of good governance desire.

Looking more closely at the model, it describes a situation which is highly conflictual. The advantage, from Kelsall's and Booth's perspective, is that it provides the means for structural change and upgrading through a top-down strategy. From the opposite ("institutionalists") perspective, it is also a situation where bribes are necessary to do business. It may be predictable in the sense that the successful firm gets a stable situation and others know to stay away. It is much like doing business with the mafia, which is a kind of stable order too.

Where the model scores low is in the general business climate; it will likely not open up for small and medium-sized enterprises (SMEs) or for competition more generally ("trial and error"). It misses a central characteristic of the market economy as an ideal, being an experimentally oriented economy. A developmental state (or Mafioso state) might be good at copying but most likely not at fostering an entrepreneurial spirit.

Secondary problems relate to the bureaucracy. Civil servants will work for the boss rather than as a professional Weberian bureaucracy. This differs from the developmental state in Asia, where an important characteristic was the "embedded autonomy" of the bureaucracy – i.e. its ability to make "rational" plans through negotiations with individual firms (noted earlier; Evans 1995). The bureaucracy acted like professional investors, not like the mafia.

Developmental patrimonialism reduces the capacity of the bureaucracy to reform the economy. This could furthermore have an indirect impact on trust in society, if trust depends on a professional bureaucracy. A key argument from the quality of government perspective is that the government sets the standard for society in general. If you can't trust the government, you will not trust other citizens either (Rothstein 2011).

Another type of problem relates to the political system. From the perspective of good governance, it is a problem in itself that the rule of law and the democratic system are not fully implemented, and it is partly a problem for the idea of developmental patrimonialism too. There will be political tensions if the leaders don't allow debate. It will be difficult to develop political parties, interest groups and independent centres of power if there are few political rights.

If the leader can't be held accountable through a political process, the risks of malpractice will increase. The leader will have more power in combination with higher risks of making mistakes. Good governance (the rule of law) puts a limit on what governments can do, which its proponents see as good for the respect for human rights as well as for the quality of decisions made.

To summarize, there are problems with developmental patrimonialism as an economic model as well as a political mechanism to support economic development.

Problems with good governance

In fairness, one should look for problems with the model of good governance too. There are many variants of this model, but for simplicity I will discuss it as a model where a proper legal system should be in place ("rule of law"), a Weberian bureaucracy should implement the rules (with room for professionals like doctors and teachers to run qualified services) and the system for designing the rules should be democratic.

One of the problems with the strategy of good governance is with the sequencing, that all elements are pursued simultaneously or in the wrong order in developing countries. The idea of good enough governance (noted earlier) was about selecting the most urgent measures first. A problem mentioned was to have multi-party competition in a corrupt society since it can lead to increased bribery by the incumbents, rather than transparency and accountability, as in an ideal situation.

Rule of law means to have a government, which is a problem in failed states. Good governance is to break up the rule of clans and force them under a common legal system. If the fundamental problem is rivalry among groups, a federal solution may be workable. In reality, this has proved to be very difficult for the proponents of good governance (Weiner 2013). Developmental patrimonialism could instead be to work with the local leaders or to elevate one of them into national command (like Gadhafi of Libya or Hussein of Iraq), which is a temporary solution, maintaining tensions under the surface.

A common problem with bureaucracies in poor countries is that they are underpaid and become dependent on bribes. Uniforms and offices can become means for individuals to enrich themselves in a desperate situation. Decent salaries and meritocratic recruitment are obvious solutions, but they depend on the collection of taxes and, in turn, on some trust in the government.

Well-functioning democracies are hard to establish too, especially where societies are deeply divided along ethnic lines. It is more beneficial with conflicts over material interests and ideas, as created through industrialization. Democracies need several parties to have lively debates. They also need media and a certain level of education and so on.

It is an empirical question to find the best sequencing, but if we take Sweden as a typical northern European country, the sequence was in the optimal order. The rule of law began to develop before Sweden was a unified country under a king. The kings and parliament struggled for supremacy but there was, for example, a supreme court before 1800. The bureaucracy followed rules, but offices were bought and sold, which put a severe limit on its professionalism. These matters were handled in small steps from 1809 until 1866, about 50 years before Sweden became a democracy. Modernization was carried out by parliament and kings. Democracy came later.

Sweden illustrates one of the points made earlier, that all good governance doesn't have to come at the same time. Non-democratic states in Europe were rather efficient and professional under their kings, due to their need to raise money to fight wars (Tilly 1992). This is seldom the case with developing countries. Colonial rulers often imposed a bad version of their own political systems, where economic production was organized to serve the need of the colonizer. Frederick Cooper (2002) has coined the term "gatekeeper state" to describe the economic model. Hence, one can question the design of economic policies as well as the design of the state. After liberation in the '60s, the former colonies in Africa had to find new economic models while at the same time designing a rational state (rule of law, bureaucracy) and launching a democratic political system.

A value conflict

All the aforementioned are related to the empirical side of the debate that developmental patrimonialism claims to be a more efficient strategy for economic growth and, secondly, for the development of good governance. These are causal claims which can be tested empirically.

The other side of the matter is normative about values to promote. It seems that developmental patrimonialism tells us to put our values on hold – i.e. that we need to postpone them in the short run to achieve them in the long run (a causal claim). The question here is if it is reasonable to overlook values such as equal rights before the law. This part of the argument isn't very clearly developed by Kelsall and Booth, but they seem to accept many exceptions from what is held dear in the global north, especially by saying that good governance has to be weighed against the problem of poverty. It is a tough message to say that people should live without the rule of law, accepting a dictatorship.

Chang (2003b) sees a role for some regulation relating to a market economy, which seems reasonable in developmental patrimonialism too. The exact contents are open for discussion. Some aspects of a legal system may be compatible with developmental patrimonialism, but at least the economic sphere must be controlled by the leader. The right to start a business and to challenge existing businesses must be limited. Authoritarian regimes and their lack of human rights come to mind.

It is difficult to argue that people should live with a bureaucracy which is loyal to the boss (the party?) rather than independently professional. It is not exactly clear what developmental patrimonialism demands, but it seems to put a limit on the professionalism of the bureaucracy. This means that citizens are not being treated equally under the law. It also means that recruitment is based on other principles than meritocracy.

It is a tough message to say that people should live without democracy, without the right to express opinions and have a say in how a country is run. Again, there may be several options under developmental patrimonialism, but at least accountability has to be limited.

Furthermore, it seems logical to argue from the perspective of developmental patrimonialism that existing practices in many countries have to be abolished. The implication of the aforementioned is that the fight against corruption should be put on hold. Civil servants who work to uphold the law should stop doing so. This is one of the most challenging claims implicit in the developmental patrimonialism, that it seems to ask for the abolishment of some of the efforts to make democracy and the rule of law work. It seems to ask for a reversal of the British constitutional tradition in developing countries.

Conclusions

Kelsall and Booth have some interesting points about weaknesses in the general emphasis on global governance, especially the strong version practiced under the infamous Washington Consensus. A valuable contribution by developmental patrimonialism is that it points to the weaknesses of global governance, especially the implicit assumption that one model fits all. It also points to interesting cases which broaden the debate on what causes economic growth.

At the same time, it needs to be pointed out that there are risks and drawbacks with their alternative strategy of developmental patrimonialism. The problems with a benevolent economic dictatorship are especially high in our times, when economies and politics are integrated on a global scale. Economic integration presents an alternative in the form of global value chains, where export could provide a fast track to economic development. In politics, it also seems risky to promote undemocratic leadership, with ever-increasing global agreements on climate change and poverty reduction, etc. It is important that global governance includes a link to ordinary citizens.

Developmental patrimonialism encourages us to think more about the problems with democracy promotion. Some of the problems seem to have their root causes in background conditions which are different from what they were in the global north. One such factor is the more ethnically divided societies, especially in Africa. Another factor is the importance in Europe of establishing the rule of law and a professional bureaucracy long before democracy was adopted. The situation is very different when all three are introduced in parallel.

Developmental patrimonialism also encourages us to think about barriers for economic development, especially that they can be political in character. The bad forms of corruption create problems in many countries and need to be addressed, as they are in the SDGs. It also encourages us to think about the coevolution of the political system with the economic system. Ending on a positive note, one may hope that increased trade also means an inflow of ideas and political pressure for further development.

References

Booth, David & Cammack, Diana 2013: *Governance for Development in Africa: Solving Collective Action Problems*, London: Zed Books.

Chang, Ha-Joon 2003a: *Globalisation, Economic Development and the Role of the State*, London: Zed Books.

Chang, Ha-Joon 2003b: *Kicking Away the Ladder: Development Strategy in Historical Perspective*, London: Anthem Press.

Cooper, Frederick 2002: *Africa since 1940: The Past of the Present*, Cambridge: Cambridge University Press.

Evans, Peter 1995: *Embedded Autonomy: States & Industrial Transformation*, Princeton: Princeton University Press.

Glenn, John 2007: *Globalization: North-South Perspectives*, Abingdon: Routledge.

Grindle, Merilee S. 2004: "Good Enough Governance: Poverty Reduction and Reform in Developing Countries", *Governance: An International Journal of Policy, Administration, and Institutions*, vol. 17, no. 4, pp. 525–548.

Grindle, Merilee S. 2007: "Good Enough Governance Revisited", *Development Policy Review*, vol. 25, no. 5, pp. 553–574.

Hall, Peter A. & Soskice, David (eds.) 2001: *Varieties of Capitalism: The Institutional Foundations of Comparative Advantage*, Oxford: Oxford University Press.

Kelsall, Tim 2011: "Rethinking the Relationship between Neo-Patrimonialism and Economic Development in Africa", *IDS Bulletin*, vol. 42, no. 2, March, pp. 76–87.

Kelsall, Tim 2013: *Business, Politics and the State in Africa: Challenging the Orthodoxies on Growth and Transformation*, London: Zed Books.

Kelsall, Tim, Booth, David, Cammack, Diana & Golooba-Mutebi, Frederick 2010: *Developmental Patrimonialism? Questioning the Orthodoxy on Political Governance and Economic Progress in Africa*, Working Paper no 9, Africa Power and Politics programme (APPP), London: Overseas Development Institute, ODI.

Landman, Todd 2013: *Issues and Methods in Comparative Politics: An Introduction*, Third edition, Abingdon: Routledge.

Moore, Mick 2001: "Understanding Variations in Political Systems in Developing Countries: A Practical Framework", Unpublished.

Rothstein, Bo 2011: *The Quality of Government: Corruption, Social Trust, and Inequality in International Perspective*, Chicago: University of Chicago Press.

Serra, Narcis & Stiglitz, Joseph E. (eds.) 2008: *The Washington Consensus Reconsidered: Towards a New Global Governance*, Oxford: Oxford University Press.

Tilly, Charles 1990: *Coercion, Capital, and European States, AD 990–1992*, Oxford: Blackwell.

Weiner, Mark S. 2013: *The Rule of the Clan*, New York: Farrar, Strauss & Giroux.

Williams, David 2012: *International Development and Global Politics: History, Theory and Practice*, Abingdon: Routledge.

Woo-Cumings, Meredith (ed.) 1999: *The Developmental State*, Ithaca: Cornell University Press.

7 National strategies of implementation

Tobias Ogweno[1]

In this chapter, we look at how governments and stakeholders are implementing the SDGs three years into their adoption. To understand the dynamics and what the stakes are, we first examine how the SDGs are different from the MDGs. Thereafter, we survey some of the approaches, plans and methodologies applied by member states and the UN in helping advance implementation of the SDGs. We also look at how the SDG approaches are different from those applied during the MDGs era and evaluate their strengths and weakness. This is intended to provide some lens into whether the SDGs have the reasonable positive effect required to put the world into a sustainable pathway.

Effective governance arrangements are a key spike in the cogwheels that drive the operationalization of the SDGs. The very elaborate nature of the UN Agenda 2030 and the SDGs require the executive, the legislature and the judicial branches of government to work in a mutually reinforcing fashion to realize the broader visions and aspirations of the goals. Therefore, without going too much into detail, we survey how some governments are going about working as a unit to promote the required synergies and to reduce the organizational barriers ("silos"), and whether these approaches are finding traction or not.

We also explore how governments are using the SDGs to manage the tensions from the desire to sustain economic growth as a requirement for meeting social needs while protecting the environment. In the end, we survey the popularity of the SDGs and give our own perspectives on how member states can enhance it as a contribution towards effective implementation of the SDGs moving forward.

What is new with the SDGs?

Since their approval, the SDGs have become the slogan for the development community. Every serious development stakeholder, whether in government, the private sector or the corporate world now have some kind of ideas about the SDGs. Many also tend to seek how they can collaborate to ensure that their resources or businesses are realigned to meet the targets and indicators of the SDGs. This was hardly the case with the MDGs, which lacked the spark and the enthusiasm of all stakeholders.

First, the 8 MDGs provided the foundation for the establishment of the 17 SDGs. It appears that having experimented with the MDGs for 15 years, member states were somehow convinced that the SDGs too are attainable within the same period despite their sheer number and targets. However, the sets of goals are different in numbers and formulation processes.

The UN, perhaps inspired by the need to save the Millennium Declaration (UN 2000) from obscurity, created the eight MDGs and handed them over to member states for adoption. This top-down approach did not benefit from a wealth of knowledge and experience of the various stakeholders who encounter development challenges on a day-to-day basis. It also occasioned lack of ownership by stakeholders.

The case was different with the 17 SDGs. From the inception of the negotiation to the adoption of the outcome, the process was one of the most transparent, democratic and innovative intergovernmental processes in recent times. Throughout its 14 interactive sessions, the OWG heard and incorporated inputs from governments and non-governmental stakeholders. At the end of it all, the process created a sense of ownership as the goals accommodated the interests of various groups in one way or the other.

Second, the MDGs were narrow and sometimes focussed on specific numbers in solving some of the most pressing global challenges. For example, it called on the development community to reduce by half the number of people suffering extreme poverty as well as to improve the lives of 100 million slum dwellers. On the converse, the SDG's call is comprehensive and universal: to end poverty in all its dimensions everywhere. The difference here is that the MDG's vision for a world free of poverty was incremental and driven by the willingness of rich nations to help realize the vision.

The SDG vision is transformative and driven by a holistic whole-of-society approach, hence the rallying call "leave no one behind". The MDGs were fashioned in the context of developed nations providing financial and other forms of support to developing counterparts, an arrangement which did not yield as much, as many developing countries still lagged behind in achievement of all the MDGs, even after their expiry in 2015. Aware of this challenge and gap, the international community integrated the incomplete MDGs into the SDG framework. Besides, the SDGs encourage all states, poor and rich alike, to mobilize resources from all sources and to individually and collectively take action.

Third, the MDGs were simple yet convoluted. The drafters lumped issues together, even those without a direct and straightforward relationship. For example, MDG goal 1 placed poverty and hunger together, which could give a lopsided impression that the fight against poverty goes hand in hand with ending hunger. The SDGs recognized such opacities and treated issues with more clarity and merit. For example, unlike the MDGs, the SDGs provided stand-alone space for poverty eradication and food and nutrition security.

The MDGs were also modest in the sense that they primarily focussed on addressing what we call the symptoms of the problem, unlike the SDGs, which focus both on providing solutions and addressing the root causes of the problems. For example, in the case of natural disasters such as drought or floods, the MDGs tackled them as they occurred by simply mobilizing the humanitarian community to provide temporary support to the victims. That was as far as it would go. The world would then wait to mobilize in case another similar challenge occurred.

The point here is that, largely, the MDGs were not anticipatory. Consequently, they failed to recommend establishment or strengthening of long-lasting structures or mechanisms for problem-solving. In contrast, the SDGs adopted a different model of isolating and giving the issues their own space in the agenda. Moreover, the SDG framework has embraced an ambition that goes beyond temporary solutions to pitching a case for actions that address the root causes of the problems and promote resilience building.

Fourth, the MDGs did not expressly recognize peace and good governance as necessary preconditions for economic stability and social development. However, despite this omission, the UN supported the work of its peacebuilding commissions through mobilization of resources and the global development community in general to support peacebuilding efforts and initiatives. Essentially, durable peace, or lack of it, relates closely to the way countries govern themselves and, by extension, it relates to sovereignty. This explains why a great majority of countries, especially among the developing, initially opposed the inclusion of the peace, security and governance issues in the SDG framework. However, in the spirit of consensus and give and take, member states finally incorporated governance and peacebuilding in the SDGs.

Fifth, the MDGs did not expressly call for the monitoring, evaluation and accountability of what participating states, donors and recipients alike were doing. That possibly explains why there was chilled implementation of some goals and why the donor community did not fulfil their official development assistance (ODA) targets. Besides, governments participated in the Annual Ministerial Review of the MDGs under the UN Economic and Social Council (ECOSOC), but there were no concrete follow-up actions on the recommendations. Whereas most governments are typically less comfortable with open and all-inclusive reviews, and prefer to disseminate only the information they feel is shareable, this area needs to be addressed to encourage transparency and accountability, especially in the management of public affairs.

The SDG framework has incorporated the sharing of quality information. Suffice it to say, one of the targets reads "by 2020 increase significantly the availability of high-quality, timely and reliable data disaggregated by income, gender, age, race, ethnicity, migratory status, disability, geographic location and other characteristics relevant in national contexts" (UN 2015). This will enable governments and key stakeholders plus potential financiers to know the facts about what they are funding as reflected on the ground and to direct resources appropriately. We refer to this information management approach as a data revolution and consider it one of the key selling points of the SDGs.

Sixth, the MDGs emphasized quantity rather than quality of the services. For example, in the case of education for all, resources were funnelled to facilitate higher enrolment for pupils and students without proportionate provision of the teaching staff and aid. Higher school enrolments without commensurate support structures eventually contributed to lowering the quality of education in some countries and negated the very purpose of education.

The SDGs altered this rationale. SDG goal 4, target 4.7 reads

> by 2030 ensure all learners acquire knowledge and skills needed to promote sustainable development, including among others through education for sustainable development and sustainable lifestyles, human rights, gender equality, promotion of a culture of peace and non-violence, global citizenship and appreciation of cultural diversity and of culture's contribution to sustainable development.
>
> (UN 2015)

By taking such a comprehensive approach, focussing both on the quantity and quality of education and its outcomes, this SDG target, if fully implemented, could play a substantial role in achieving a more economically prosperous, socially equitable and environmentally conscious world.

The context of the SDGs

Member states negotiated the SDGs with the full benefit of retrospection of the MDGs strengths and weaknesses, and the vision of a sustainable present and future world. They were aware of most of the shortcomings of the MDGs, the prevailing global economic realities and the deteriorating state of the environment. Therefore, they were determined to mend the gaps that made the MDGs less impactful while building on the strong areas of the goals.

Also, most negotiators of the SDGs were familiar with the fact that the 2008–09 global economic recession had severely affected many traditional donor countries and that many were still recovering from the contraction. Accordingly, they knew only too well that in the post-MDG era, the ODA would form an insignificant part of financing for development and that if the SDGs are to succeed, a more sustainable financing model had to be thought through.

Besides, some developed countries had put up a strong case that the SDG process needed to revisit the global funding terrain, which unrealistically continued to place a heavy burden on them. On the face of it, this argument holds some truth, as many developing countries had recorded impressive economic growth and maintained an upward trajectory, even during the economic meltdown. The new and emerging economies of Brazil, Russia, India, China and South Africa, among others, had secured space in the global community as new economic powerhouses. Therefore, the feeling among many developed countries was that this group of states and others that were doing comparatively well should not only place the ODA card on the table but also take commensurate international obligations and responsibilities.

A question is if this argument can support the representation that SDGs have played down the donor-recipient design and encompassed a more practical and radical approach to sustainable development. In our view, the answer is yes. SDG goal 8 – "promote sustained, inclusive and sustainable economic growth, full and productive employment and decent work for all" – was agreed as a win-win strategy to ease the burden on traditional donors by encouraging governments, especially of the developing world, to rely mostly on their internally generated resources to fund their own sustainable development.

Therefore, all member states must enhance their own revenue-generation capacities and fund a substantial part of their own development, with the only exception being the least developed countries, which qualify for some level of support. This is not to say that rich nations should not support their developing counterparts. The decision as to whether to support or not is an entirely sovereign one. What is clear is that the fulfilment of the SDGs requires all governments and stakeholders to embrace new and innovative implementation approaches, especially with regard to mobilizing resources and deploying strategies that "leave no one behind".

The international community has adopted several agreements in the area of sustainable development after the UN Conference on Sustainable

Development, also known as the Rio+20 Summit, held in 2012. In addition to the SDGs are the Addis Ababa Agenda for financing and the Paris Agreement on climate change. The governments of the world negotiated these agreements in different tracks and on the merit of the issues they represent. They relate in one way or other and their implementation can galvanize some kind of synergy required especially for the SDGs.

Besides, the SDGs have been naturally favoured by other factors. The expiry of the MDGs created a significant goals gap. The experience of the world working around the MDGs as some sort of global goals was still fresh, and the dwindling natural resources vis-à-vis the growing world population and environmental footprint demanded transitioning to sustainability. The existence of a critical mass of developing states with the potential for real transformation is a plus for the SDG stakeholders. These factors are the four strategic advantages that make the SDGs the "lucky bird with wings to fly". These factors should help stimulate further ambition for the implementation of the SDGs.

The Voluntary National Reviews

Since their rollout in January 2016, member states and the development community have continued to demonstrate substantial interest and commitment to the realization of the SDGs. The last three years (2016, 2017 and 2018) have witnessed 112 countries present a total of 111 Voluntary National Reviews (VNRs) to the sessions of the High-Level Political Forum (HLPF) convening under the auspices of the UN ECOSOC in New York.

We note that due to varied interpretations, the submissions have their own limitations, as a great number of states did not adhere to the reporting format prescribed by the Department of Economic and Social Affairs of the UN (UN DESA). The reporting on the material content also lacked homogeneity as countries reported on what they wanted. This has made it challenging to compare the submissions in a meaningful manner.

The HLPF is the UN central platform for the follow-up and review of the UN 2030 Agenda for Sustainable Development and the SDGs. The forum is an avenue for the full participation of all members states of the UN and members of specialized agencies, enabling countries to assess and compare on progress, achievements, challenges and gaps they are facing while effecting the UN 2030 Agenda and the SDGs.

The HLPF convened the first SDGs review session in July 2016. According to the UN Department of Economic and Social Affairs, the 22 countries that participated in that session's VNRs (Figure 7.1) mostly reflected on the challenges, gaps, achievements and lessons learnt since the start of the SDGs implementation (UN 2016). The synthesis report states that many

	2016 (22 countries)	2017 (43 countries)	2018 (47 countries)
Countries	China	Afghanistan	Albania
	Colombia	Argentina	Andorra
	Egypt	Azerbaijan	Armenia
	Estonia	Bangladesh	Australia
	Finland	Belarus	Bahamas
	France	Belgium	Bahrain
	Georgia	Belize	Benin
	Germany	Benin	Bhutan
	Madagascar	Botswana	Cabo Verde
	Mexico	Brazil	Canada
	Montenegro	Chile	Colombia
	Morocco	Costa Rica	Dominican Republic
	Norway	Cyprus	Ecuador
	Philippines	Czech Republic	Egypt
	Republic of Korea	Denmark	Greece
	Samoa	El Salvador	Guinea
	Sierra Leone	Ethiopia	Hungary
	Switzerland	Guatemala	Iceland
	Togo	Honduras	Ireland
	Turkey	India	Jamaica
	Uganda	Indonesia	Kiribati
	Venezuela	Italy	Lao People's
		Japan	Democratic Republic
		Jordan	Latvia
		Kenya	Lebanon
		Luxembourg	Lithuania
		Malaysia	Mali
		Maldives	Malta
		Monaco	Mexico
		Nepal	Namibia
		The Netherlands	Niger
		Nigeria	Paraguay
		Panama	Poland
		Peru	Qatar
		Portugal	Romania
		Qatar	Saudi Arabia
		Slovenia	Senegal
		Sweden	Singapore
		Tajikistan	Slovakia
		Thailand	Spain
		Togo	Sri Lanka
		Uruguay	State of Palestine
		Zimbabwe	Sudan
			Switzerland
			Togo
			United Arab Emirates
			Uruguay
			Vietnam

Figure 7.1 The countries that participated in the VNRs in 2016–18

countries emphasized the importance of involving various stakeholders in situating priority areas, implementation and review; the critical roles of good governance, financing, effective international support, technology and innovation, capacity building and trade in delivering the SDGs (ibid.).

According to the synthesis report, developing countries see the principle of common but differentiated responsibilities still relevant even for the economies in transition and middle-income countries. The report further notes that countries had specified some of the main challenges raising as the following: lack of ownership of the SDGs, coordination of various government sectors, financial and human capacity limitations, weak capacity in production of reliable data, and the establishment of inclusive mechanisms for follow-up and review. The report also notes that countries took varied approaches with regard to the specific SDGs, which makes it difficult to deduce a common fair view of the status of implementation (ibid.).

In 2017, 43 countries, almost twice the number in the previous year, participated in the VNRs. This time round, the session reviewed implementation of several SDGs and a greater majority of countries gave specifics rather than generalities in their SDGs implementation reviews. Although this was an improvement from the preceding year, we have been unable to compare country-by-country implementation progress because there is a lack of a common standard. In addition, "no uniform way of reporting on SDG specific implementation in the VNRs exists, and countries chose numerous different methods depending on their national circumstances" (UN 2017).

However, many countries reported progress in establishing or strengthening existing formal structures such as inter-ministerial coordinating offices, committees, or commissions to support coherence, integration, coordination and multi-sectorial involvement. Some countries, such as Sri Lanka and Mali reported having dedicated parliamentary committees on Agenda 2030 and the SDGs. In some countries members of parliament were even included in the official delegation to the 2017 HLPF session (ibid.).

Despite the milestones, several countries highlighted the following challenges and called for enhanced efforts to address them: policy incoherence, weak capacity of national statistical institutions to perform their tasks, poor coordination between the different levels of government, mainstreaming the SDGs into the policies and programmes of all relevant line ministries and lack of a "whole-of-society" approach to make the SDGs truly national endeavour (ibid.).

The HLPF session 2018 considered presentations by 47 countries and appraised the implementation of SDG goal 6 (clean water and sanitation), goal 7 (affordable and clean energy), goal 11 (sustainable cities and communities), goal 12 (responsible consumption and production) and goal 15 (life on land).

Over the last VNRs, many governments have highlighted the following key messages. There is commitment to promote the integrated and indivisible nature of the SDGs. However, many countries face weak policy coherence and multi-sectorial coordination which undermine these efforts. There is a need to scale up awareness-raising including through organizing conferences, workshops and festivals among others to increase the profile, visibility and ownership of the SDGs. The feeble capacity or the lack of it altogether in many developing countries remains a major hindrance to the implementation of the SDGs as affected countries are unable to undertake cost analyses and needs assessment to support resource mobilization. Effective stakeholder engagement remains elusive in some countries due to lack of adequate finances for organizing and maintaining structured with stakeholders (UN 2018).

By the time of publishing this chapter, 51 countries had volunteered for presentation of VNRs in the 2019 session of the HLPF, ten of which will be carrying out their second or even third VNR. It would review the implementation of goals 4 (quality education), 8 (decent work and economic growth), 10 (reduced inequalities), 13 (climate action) and 16 (peace, justice and strong institutions). In addition, the forum will also consider the means of implementation (goal 17) at the session.

National strategies: different approaches, same destiny

Agenda 2030 and the SDGs are famous for their lack of a collective and universally agreed implementation strategy. This is in line with the fact that countries apply different approaches relevant to their national circumstances.

The UN DESA defines a national sustainable development strategy as "a coordinated, participatory and iterative process of thoughts and actions to achieve economic, environmental and social objectives in a balanced and integrative manner" (UN DESA undated). It explains that "national sustainable development strategies are based on a number of principles: the principles of country ownership and commitment, integrated economic, social and environmental policy across sectors, territories and generations, broad participation and effective partnerships, development of the necessary capacity and enabling environment as well as a focus on outcomes and means of implementation" (ibid.).

As mentioned earlier, member states did not include an implementation strategy for Agenda 2030 and the SDGs and left the issue for each state to decide. Because of this gap, the information provided in the VNRs generally demonstrates that countries are applying different national approaches, plans, strategies and methodologies in their efforts to achieve the SDGs.

However, approaches in some sectors appear to be similar across the vast majority of states. For example, according to the submissions in the various

VNRs, many national strategies for the SDGs have identified coordination, participation and actions stand out as the key functions driven by the principle of national ownership, integration, participation, capacity building, outcomes and implementations. With this information, we can safely argue that countries know what they need to do and how they will do it to arrive at the same answer: "different approaches, same destiny".

Stakeholder engagement

Many countries that have participated in the VNRs have reported embracing multi-stakeholder participation in the planning and review of the SDGs. For example, Kenya, Ethiopia, Jordan, the Netherlands and Slovenia reported involving women, youth, people living with disabilities, academia and members of the private sector, among other stakeholders. This is crucial because the success of the SDGs will depend on how the segments of society are treated. There is evidence that when stakeholders are included in a process, it creates greater awareness, ownership and appreciation among stakeholders, which are important for driving success of the SDGs.

As highlighted previously, one of the distinguishing features of the SDGs is that the goals tend to treat the three pillars of sustainable development (economic, social and environment) in a holistic, integrated and balanced manner. The "means of implementation" has a special place with 42 targets (with letters) across all the goals focussed on it and the final goal 17 exclusively devoted to the same. The UN categorizes the means of implementation to include finance, technology, capacity building, trade, policy coherence, partnerships and, finally, data, monitoring and accountability.

However, no implementation targets address themselves to interlinkages and interdependencies among goals. This makes integration difficult, as policy designers and implementers may not have the benefit of cross-fertilization of ideas. We believe governments must correct this anomaly through an all-inclusive approach to implement the SDGs successfully. In addition, integration must be visible, practical and real across sectors (e.g. finance and energy), across societal actors (local authorities, government agencies, private sector and civil society) and between and among countries.

Integrated governance

The global SDGs recognize that no single institution can effectively tackle sustainable development challenges. All actors in public and private spheres have a critical role to play. The UN Agenda 2030 stresses the importance of establishing a strengthened institutional framework for sustainable development at all levels to promote the integration of the economic, social and environmental pillars of sustainable development (UN 2015).

Owing to the fact that countries have different sets of institutions, levels of development, priorities and circumstances, the goals encourage member states to adopt governance arrangements or mechanisms that can work in a coherent and coordinated manner to promote synergies and reduce sectorial barriers ("silos"). Member states are responding to this call by either setting up new institutional mechanisms or strengthening existing ones for policy integration, coordination and accountability of government agencies, stakeholder participation, monitoring and evaluation and reporting.

Through inclusive and vibrant institutions, governments and stakeholders can prioritize the implementation of the goals and encourage the participation of various actors within and outside of the government. This requires dynamic and systematic coordination to forestall possible challenges, such as competition, duplication of roles, misalignment of resources and policy misapplication, among others. However, coordinating efforts across the goals and recognizing the interactions between them are key challenges.

The VNRs have established that domestication or nationalization of the SDGs to attune the goals to various national needs, priorities and circumstances stands out in almost all the cases as a top priority for most governments. Several countries have established special forms of institutions. Lao Peoples' Democratic Republic, for example, has established a National Steering Committee for SDGs implementation chaired by the prime minister. The committee also includes members from all relevant ministries and state agencies, and nearly 60 per cent of its Eighth National Social-Economic Development Plan indicators are linked to SDG indicators.

Countries such as Colombia and Sweden have established cross-ministerial committees with the purpose of mapping out a national strategy for the interacting goals. In the same breath, the National Parliament of Timor-Leste has also passed resolution number 34/2015, which recognizes and commits to the 17 SDGs.

Generally, parliaments as representatives of the people have central roles to play in the implementation of the UN Agenda 2030 and the SDGs through enacting enabling legislations, providing appropriate budgetary support and oversight function and ensuring social inclusion. In Pakistan, for example, the parliament has adopted the SDGs as part of its National Agenda, examining how inclusive their processes are.

Legitimate, innovative and flexible governance

Most of those who submitted their VNRs informed that they had put in place governance arrangements to promote the implementation of the

SDGs, including those that encourage effective engagement with multi-stakeholders. Therefore, there is a need for political goodwill and commitment from all for such arrangements to succeed. Suffice it to say, coordinating entities, especially those involving a broad spectrum of actors in government and outside, need legitimate, innovative and flexible mechanisms backed with real political clout and influence. This is necessary to help mobilize people and institutions for collective action across different sectors, levels and scales. This is especially crucial given that players may have conflicting interests and that it sometimes can be difficult to strike win-win solutions.

Strong and legitimate institutions can help make the difference by focussing on equity, justice and fairness plus the mechanisms for follow-up and review to ensure accountability for resources and outcomes, especially where difficult trade-offs are involved.

Others agree that the success of the SDGs needs effective governance arrangements and go even further to explicate that the governance arrangements should be strong enough to promote collective action across multiple sectors and scales, to make difficult trade-offs with specific focus on equity, justice and fairness, as well as to ensure mechanisms that hold societal actors to account regarding decision-making, investment, action and outcomes (Bowen et al. 2017).

Agenda 2030 for sustainable development stipulates, "the goals and targets will stimulate action over the next 15 years in areas of critical importance for humanity and the planet" (UN 2015). However, the agenda has not provided a corresponding global strategy for realizing this expectation. Before we undertake a full discussion, we consider it important to provide some standpoints on the likely implications of the statement and the omission.

In our assessment, the areas of critical importance for humanity and the planet alluded to in the agenda mean that the 17 SDGs and their 169 targets do not have a hierarchy of importance. All the goals and targets are intrinsically important, and a country or a development agency can choose whichever to implement, provided it contributes to the broader SDG aspirations. In that regard, the UNDP, for example, has its SDGs strategic plan focussed on supporting governments to integrate SDGs with components on poverty eradication, democratic governance and peacebuilding, climate change and disaster risk and economic inequality components into their national development plans and policies as the key areas.

Similarly, we can interpret the omission of a globally shared implementation strategy in the SDGs package as deliberate recognition that there is no one development approach that is applicable to all. The SDGs and their targets, being universal in scope and application provide all governments and relevant stakeholders with the flexibility in choosing what they consider particularly crucial to implement, when and how. Put differently, if poverty eradication, for example, is the top-most priority for country X, then it should be justifiably unobjectionable for the government of country X to reach out to willing partners in mobilizing and aligning the required resources towards that direction instead of other areas.

National development plans

Many countries have embraced the implementation of the SDGs in their national development plans. This is supported by the findings in a policy brief by AidData, confirming that "countries have already begun work to link national development plans to the SDGs, with varying approaches including localizing indicators into their own national development plans" (Aid Data 2018). Conversely, the brief warns that in some countries, the lack of or weak capacity to track spending towards national development activities and progress on indicators, and a clear methodology to see how they map to the SDG indicators, is making effective coordination difficult (ibid.). AidData further suggests that some countries, such as Rwanda, have built locally relevant indicators into their national development plans to track progress and are working to determine which of these indicators are (or are not) reflected in the SDGs (ibid.).

Managing the tensions

The first area of tension is what we describe as rivalry, pitting those championing climate actions against those for development. For centuries, the world has practiced economic models of growth that have considerably reduced poverty levels but at the same time produced heavy pollution, especially from industrial processes. For example, the world derives about 80 per cent of its energy from coal, oil and natural gas. In addition, commercial agricultural practices employ chemicals and conversion of large tracts of land into farmland. While these practices are altering the natural environment and ecosystems, the fact is that many poor countries require commercial energy to turn around their economies. However, it is cheaper to generate energy of a commercial calibre from the hydrocarbons compared to renewable energy alternatives.

Similarly, food and nutrition security are issues of concern in the developing world. With its population growth, which is higher than in the developed world, the food insecurity situation can get out of hand, unless targeted interventions are adopted to address the issue. While SDG goal 13 requires the world to "take urgent action to combat climate change", many climate adaptation experts argue that weathering the effects of a warming world will mean building resilience among the world's vulnerable people, including through promoting human development. We, therefore, find ourselves asking the question, are combatting climate change and the pursuit of sustainable development two interrelated yet separate processes, or are they two halves of an indivisible whole?

As mentioned elsewhere, the Paris Agreement on climate change and the SDGs followed two separate negotiation tracks. Most states still prefer that formal negotiations or reporting on these issues be undertaken at the Conference of the Parties to the UNFCCC for the former, and the HLPF for the latter. In our view, this should change. It is more logical for states to discuss their nationally determined contributions for climate action in the context of SDG review to promote synergies and to demonstrate how the climate instrument can contribute to achieving the indivisible whole of sustainable development. However, states prefer separate discussions as a practical way of diffusing tensions and getting their own trade-offs.

The second area of tension is the lack of clear and reliable information. The common saying "information is power" holds true here. To put it in another context, a people empowered with the right information tend to make the right choices and decisions, whether in investments, services procured or goods consumed.

Although the SDGs are more elaborate and more complex compared to the MDGs, the closer we scrutinize them, the clearer the interconnections and interrelationships begin to emerge. To promote political buy-in across various segments of the social classes, decision-making on SDG issues should be grounded on solid, accessible, digestible and relevant information. Governments can manage the tensions arising from the misinterpretation of the SDGs by balancing the knowledge on the growing complexity of socio-economic and environmental systems with the knowledge relatable to the people on the ground. For example, explaining to ocean-dependent communities that conservation of mangroves can enable them earn carbon credits while improving their fish catch and resilience of their coastal development and infrastructure can serve as an incentive to sustainably harvest the mangrove.

There is no doubt that human beings and other species have a special relationship of inter-dependence. We, for example, cultivate the crops that feed some insects. At the same time, the insects pollinate the crops contributing to food security in addition to performing other ecological functions. It, therefore, follows that destroying one system can upset or alter a whole value chain. The concept of sustainable development recognizes these interrelationships and is being promoted to maintain some kind of balancing act. However, more often than not, the economic, social and environmental needs of the human species is given priority over other species.

If the world is to achieve the SDGs, then policy and decision-makers must work around the goals to balance the three needs through synergies and trade-offs. Governments can encourage this by establishing innovative frameworks to promote the understanding of the interactions between SDG targets for supporting synergies and those for identifying policy priorities for trade-offs. This method should apply to the SDG targets that support the realization of other related targets or to those that can offer sustainable alternatives.

Citizen engagement with the goals

The intergovernmental-led negotiations that formulated the SDGs remain one of the most engaging, democratic and representative processes in recent times. The style created a sense of ownership and built considerable confidence among the negotiating parties and the constituencies they represented. We believe too that for the SDGs to succeed, stakeholders should feel a part of the implementation process and that governments should facilitate this by working with their public and private citizens to know how they feel about the goals. This is more important given that "understanding what different constituencies know and think about the SDGs is a crucial starting point" (OECD 2017).

There is large popular support for the SDGs, which implies three things. First, the world is on a good SDGs awareness trajectory, especially considering that the goals have been in existence for only three years since their adoption. Governments should work with relevant stakeholders to reach out beyond national governments and sustain awareness initiatives by providing political goodwill in support of the implementation of the goals.

Second, the fact that support for the SDGs is much stronger in emerging economies than in developed countries proves that the goals have found some traction in these countries, which ordinarily would be expected to demonstrate a lackadaisical approach towards the goals. What it means is that the SDGs can have more impact if the sustainable development community supports their effective implementation in the developing countries. This is more so because this group of states is where change of economic growth model can occur with reasonable trade-offs.

Third, the fact that the youth are more aware of the SDGs than the average person is demonstrable of their interests in the goals. Governments should take advantage of the youth as dynamic agents of change to popularize the SDGs and bring about the much-needed societal transformation.

In conclusion, the SDGs continue to find traction across governments and stakeholders.

Note

1 Tobias Ogweno is a Kenyan diplomat and researcher currently stationed at the Kenya Embassy in Rome, Italy. Previously, he served at the Second Committee of the UN General Assembly while based at the Kenya Mission to the UN in New York. He practices international relations with specific interests in sustainable development. He participated actively in negotiating the SDGs, as well as other international agreements, including "The Future We Want" at the UN Conference on Sustainable Development (Rio+20) and the New Urban Agenda.

References

Aid Data 2018: "Linking the SDGs with National Development Planning for Better Coordination", *Blogpost*, May 11. www.aiddata.org/blog/linking-the-sdgs-with-national-development-planning-for-better-coordination

Bowen, Kathryn J., Cradock-Henry, Nicholas A., Koch, Florian, Patterson, James, Häyhä, Tiina, Vogt, Jess & Barbi, Fabiana 2017: "Implementing the 'Sustainable Development Goals': Towards Addressing Three Key Governance Challenges: Collective Action, Trade-Offs, and Accountability", *Current Opinion in Environmental Sustainability*, vol. 26–27, pp. 90–96.

OECD 2017: *What People Know and Think about the Sustainable Development Goals: Selected Findings from Public Opinion Surveys*, Compiled by the OECD Development Communication Network (DevCom), June.

UN DESA (undated): *National Sustainable Development Strategies – The Global Picture*, Note prepared by the Division for Sustainable Development of the United Nations Department of Economic and Social Affairs, Informal publication.

United Nations 2000: *Millennium Declaration*, Resolution adopted by the General Assembly on 18 September 2010, A/Res/55/2.

United Nations 2015: *Transforming Our World: The 2030 Agenda for Sustainable Development*, Resolution adopted by the General Assembly on 25 September 2015, A/Res/70/1.

United Nations 2016: *Synthesis of Voluntary National Reviews*, Division of Sustainable Development, Department of Economic and Social Affairs, New York.

United Nations 2017: *Synthesis of Voluntary National Reviews*, Division of Sustainable Development, Department of Economic and Social Affairs, New York.

United Nations 2018: *Synthesis of Voluntary National Reviews*, Department of Economic and Social Affairs, New York.

8 Conclusions of the study

This chapter will summarize the findings of the previous chapters and provide answers to my initial questions. It will discuss strategies and challenges of the Sustainable Development Goals (SDGs).

Summary

This study has aimed to provide a better understanding of the SDGs, also known as Global Goals or Agenda 2030, and the challenges involved in the implementation of these goals, especially in developing countries. The book has taken a critical and constructive approach, pointing out risks as well as possible remedies. The SDGs are seen as an opportunity for a global conversation on what works in solving some fundamental problems related to poverty and environmental degradation.

The SDGs are remarkable for the global commitment on a set of ambitions to reach by 2030, but also for the lack of a strategy of implementation. The choice of appropriate action is handed over to national governments, which vary not only in terms of the challenges they face but also in terms of previous experience and resources available. The first chapter described the 17 SDGs and pointed out some of their key characteristics, like their lack of explicit strategies and their unclear relation to other global policies.

The study has taken a forward-looking perspective, treating the SDGs as a case of a more general discussion about how policies can be implemented under difficult circumstances. It is an early evaluation ("ex ante") of what the SDGs demand from national governments. Chapter 2 gave an overview of existing literature to point out the research gap this book aims to fill. It places the book within the boundaries of comparative politics, IPE, policy analysis and economics.

The book has advanced the idea that governments can improve their policies over time, especially if they see their plans as hypotheses about what works in promoting sustainable development. To do so, they need to identify drivers and barriers for sustainable development. The academic debate is focussed on the concept of goals, while this book suggests that the concept of a learning policy provides a practical way forward. A learning policy is a policy by trial and error, starting from hypotheses and revising them when evidence suggests so. This makes it necessary to think about causality – i.e. in terms of drivers and barriers for sustainable development.

In Chapter 4, an inherent strategy was identified within the 17 SDGs and their 169 targets – i.e. a statement on causal relations with drivers and barriers to achieve the goals. The strategy treats the social and environmental goals as the highest ends. The economic goals are important means to reach the other goals, together with environmental regulation and taxation. The goals point to good governance (state capacity) as a prerequisite for all the dimensions of sustainable development. Peace, justice and institutions are important prerequisites for the other goals.

The inherent strategy was compared in Chapter 5 to three political perspectives from IPE. The three stylistic perspectives of liberal, nationalist and Marxist world views gave a variety of interpretations of what sustainable development would look like. The perspectives differ, for example, in their views on states and markets as coordinating devices, and in their nationalist or internationalist perspectives. This was used to broaden the debate on the SDGs and to classify the SDGs, which turn out to be mainly a nationalist strategy.

Chapter 6 looked more closely at the key challenge to achieve good governance – i.e. an efficient administration (a state) to make decisions and implement them. The lack of a well-functioning organization may be the most important of all challenges since it makes other implementation more difficult. The study looked at a recent debate on whether good governance must come first or if it is an effect of other kinds of development. The proponents of alternative views, such as "developmental patrimonialism", contribute to a deeper understanding of the complexities in state-building and economic development. However, good governance remains an important value, as well as a foundation for a well-functioning government and society.

Chapter 7, by Tobias Ogweno, discussed what can be learnt from the national implementation of the SDGs during the first three years (2016–18), as reported through the Voluntary National Reviews (VNRs). The chapter began by showing what was new with the SDGs for the national governments. It went on to describe some findings and what they imply about the tensions facing governments. One aspect is that there is a need to further develop legitimate, innovative and flexible forms of governance.

Conclusions

My first question was if there is a way of understanding the SDGs which gives them a fair chance of achieving their goals. The answer was that there is another approach which shifts the attention from the goals to the learning over time, which opens up for a conversation on what works in various contexts. Such learning over time is only possible if there is an attempt to understand driver and barriers – i.e. to elaborate on the causal mechanisms involved.

The second question was if there is an inherent strategy in the SDGs. The answer was that there is an elaborate strategy, though some steps are not clearly indicated. The social and environmental goals are most clearly stated as ends, while the means are only briefly indicated. The economic goals are much elaborated on in terms of ends and means. They are also the means to reach the social and environmental ends. Regulation and taxation are key items, which indicate the important role played by the government. Hence, the last goals (numbers 16 and 17) are causally primary to all the goals.

The third question was to take advantage of ongoing research to spell out some more elaborate hypotheses on what needs to be done – i.e. to identify drivers and barriers for the achievement of the goals. The first part was to look at alternative views of drivers and barriers in Chapter 5. The second part was to focus more specifically on the topic of governance in Chapter 6. Chapter 5 took advantage of stylized perspectives to open up for a wider debate. Chapter 6 started from an alternative view on good governance to show some of the complexities involved.

The future

This study is the beginning of an evaluation of the SDGs rather than the final word. Its contribution is mainly in the elaboration of the goals and their intended causal mechanisms. It has opened up a discussion about what is relevant to take into consideration in the further implementation.

The next step would be to look more in detail at the national implementation, what governments do and how it all relates to the national context. The national learning should be about the specific drivers and barriers in the national context. At the global level, it would be of additional interest to compare national strategies – for example, in a typology of situations. Such a typology could identify types of focus areas and types of challenges. It would be of value to identify success and failure (or at least various kinds of success) to see what the successful national strategies have in common. That would add empirical data to the debate on what works in various contexts (Landman 2013).

Under the best of circumstances, this may not only add to our understanding of the implementation of the SDGs. It could also shed some light on the interesting hypothesis noted earlier, that environmentalism can contribute to state-building and, hence, to economic development (Death 2016). It could also add to the discussion – and the experience – of a rapid transition from failed states to the rule of law, good governance and democracy. This is something of a political trilemma (Rodrik 2011), where it is difficult to pursue all three at the same time. Should one of them be put on hold to begin with? Thirdly, it could add to our understanding of the relationship between governments and their civil societies. The quality of government hypothesis states that the government is of primary importance since it is the government, which has an impact on the level of trust in society (Rothstein 2011), whereas an older understanding is that civil society is necessary to make governments civilized and democratic (Tocqueville 2003). The relationship may depend on the particular circumstances.

Finally, it will, of course, be interesting to follow the implementation of the SDGs to see what the outcome will be and which factors turn out to be the most important for its success. The SDGs make several controversial assumptions about causal mechanisms, such as the importance of economic development in general, and trade and technological development in particular. Equally interesting is the role of increased taxation to pay for social development. This will put further pressure on governments to pursue policies which are accepted and trustworthy, in democracies as well as authoritarian states. This underlines how challenging the SDGs really are, not just in the environmental dimension.

References

Death, Carl 2016: *The Green State in Africa*, New Haven: Yale University Press.

Landman, Todd 2013: *Issues and Methods in Comparative Politics: An Introduction*, Third edition, Abingdon: Routledge.

Rodrik, Dani 2011: *The Globalization Paradox: Democracy and the Future of the World Economy*, New York: W.W. Norton & Company.

Rothstein, Bo 2011: *Quality of Government: Corruption, Social Trust, and Inequality in International Perspective*, Chicago: University of Chicago Press.

Tocqueville, Alexis 2003: *Democracy in America*, London: Penguin Classics (originally 1835 and 1840).

Index

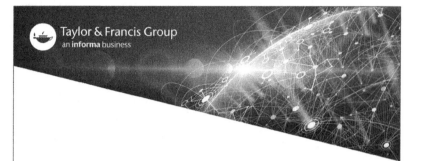

Printed in the United States
by Baker & Taylor Publisher Services